THE COMPLETE AND EASY

›BEEKEEPING‹

APPLE

THE COMPLETE AND EASY GUIDE TO

BeeKeePING

A Fascinating Reference with Recipes for Enjoying Your Produce

Kim Flottum

Published in the UK in 2005 by
Apple Press
7 Greenland Street
London NW1 0ND
United Kingdom

www.apple-press.com

Reprinted in 2005, 2006, 2007 ,2009

ISBN 978-1-84543-021-4

10 9 8 7 6 5

This publication is designed to provide useful information to readers on the subject of hobby beekeeping. The Publisher and Author have used their best efforts in preparing this book. Readers should protect themselves by following instructions on all beekeeping equipment and products before taking up this hobby. The author and publisher assume no responsibility, legal or otherwise, for an individual's injuries, damages, or losses which may be claimed or incurred as a result, directly or indirectly, of the use of the contents of this publication.

Design: tabula rasa
Main cover image: Clive Nichols Garden Pictures/www.clivenichols.com
Illustration by Michael G. Yatcko

The hive featured on the cover of this book is called a WBC Hive, named for its inventor William Boughton Carr. This hive was common in England in years past, but its popularity has waned in favor of less exotic, less complicated fare. Nowadays, the WBC Hive symbolizes a gentler time. It evokes an image of elegant estate grounds, skilled gardeners, and beekeepers tending the hives.

Printed in Singapore

⇥ CONTENTS ⇤

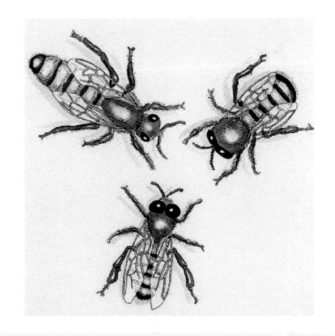

This book and the process that brought it here are dedicated to:

Chuck Koval—Who let me in;
Eric Erickson—Who let me learn;
John Root—Who let me use it all;
and Kathy Summers—Who brought it all together.

And, of course, Nancy and Buzz; Jeanne and Katie; Jim and Dawn; Mom, Dad, Jessica, Tom, Bob, Julie, and Susan; Richard, Roger, Clarence, Mark, Malcolm, Steve, James E., and Ann; and certainly the thousands and thousands of beekeepers I've been privileged to meet and who have shared what they know with all of us. And, the bees.

Preface

Pretty much every book on keeping bees is the same because beekeeping books haven't changed much in the last 150 years. The authors say the same things, do the same things, and show the same things.

This book breaks that tradition, challenges the conventional wisdom, and brings hobby beekeeping into the twenty-first century. The premise here is *not* how cheaply you can do it but how *well* you can do it; *not* what you can get by with but how much you can accomplish; *not* how much profit you can take but how much fun you can have; *not* how has it been done for the past 150 years but how it should be done today.

Here, you'll learn how to take advantage of the absolute latest in equipment technology, use the most recent findings in honey bee biology to understand your bees and succeed as a hobby beekeeper, and discover management concepts never before published in a beekeeping book.

If you follow the lead of this book, you will save precious time without sacrificing the quality of your hobby. That's guaranteed. You also won't have a bad back, way too much honey, or angry neighbors. What you *will* have is enough time to enjoy your bees, share your bounty with friends and family, and keep doing the things in the rest of your life that you enjoy.

Be warned though: Keeping bees can be addictive, and there's no known cure. But then, no one has ever looked for a cure. No one has wanted one.

Above all, enjoy the art and the science of having honey bees in your garden. Join a beekeepers' club, read the journals, burn beeswax candles, and eat honey every day. Smooth on a healing lotion made with honey and beeswax, and benefit from great garden produce pollinated by bees. What, I ask you, could be better?

Kim Flottum

→ INTRODUCTION ←

aving bees in your backyard is a good idea. Honey bees are necessary for pollinating plants to ensure a better fruit set and bigger crops. More and more people are discovering the joys and advantages to keeping bees in their yards. In fact, you may even know somebody who has bees.

In your busy life, tending your garden, fruit trees, and even your lawn are some of the things that you not only make time for but that you also enjoy. Being outdoors in the sunshine, with grass under your feet, feeling the warm soil and a gentle breeze is hard to beat, and having honey bees buzzing around fits right in.

With garden harvests a part of your life, cooking up simple dishes using your bounty is probably already second nature. Adding bees to your routine and adding your own honey to the table will allow you to reap what you sow all year long.

But where do you start? What do you need? And, most important, how much time will it take?

If you're like me—and like most other people today—time is important. So, how much time does it take to set up and take care of a couple of colonies of bees? Tending bees is a lot like taking care of a garden. There's a flurry of activity in spring, maintenance in summer, and harvest in fall. Over a season, your bees will take a bit more of your time than you spend caring for your cat, but less time than you spend with your dog.

Like any new activity, there's a learning curve in beekeeping, so the first season or two will require more of your attention than will be needed once you have some experience under your belt. And like a garden, there's prep work before you begin each new season and some equipment you'll need to get started.

Oh, and bees do sting. Let's get this right out front: They aren't out to get you, but they will protect themselves when disturbed. But think for a moment—bramble thorns scratch, mosquitoes bite, and yellow jackets are just plain nasty. Cats and dogs also scratch and bite; it's the way things are, plain and simple. But you wear gloves to prune your rose bushes, you wear mosquito repellant when outside at dusk, and if you don't tease your pets, they probably don't give you too much grief. It's the same with your bees. Work *with* them, use the tools you have for good management, and wear the right gear. Even when using gloves and long sleeves, stings happen, but if you are smart and prepared, they will be rare events. When brambles, mosquitoes, cats, dogs, the scratchy stems of zucchini plants, or honey bees cause that momentary ouch, figure out what you did to cause the ouch, utter a soft curse, rub the spot, and move on.

At the beginning of the twentieth century, the A. I. Root Company in Ohio was the world's leading manufacturer of beekeeping equipment.

So, if having a couple colonies of honey bees out back sounds like a good idea because you want a better garden, more fruit, honey in the kitchen, maybe some beeswax candles, skin creams, and other cosmetics for the bath, let's find out what thousands of beekeepers already know.

A New Concept

Just a few years ago, after a century and a half of very little change, there was a revolution in how beekeeping equipment is produced. This was the result of the technologies of woodworking, plastic molding, and automation coming together, finally, for equipment manufacturers. If time is critical in your lifestyle, this revolution is important, so let me explain.

Manufacturing beekeeping equipment hit its stride in the early twentieth century. Innovations including electricity, rail transport, and rapid communication came together, allowing manufacturers to cheaply mass-produce the necessary items, to reliably transport them to customers, and to advertise their products to most of the population. The heyday of modern beekeeping in the United States had arrived. But the U.S. still had a rural economy, and cost was a limiting factor for the customer. Manufacturers realized this, and the essentials of the equipment needed—wooden boxes (called *supers*) and their components (called *frames*)—were made as knock-down kits so that the customer could put them together when they arrived. The labor costs were shouldered by the end user, who spent time, rather than money, assembling all those pieces. Moreover, if additional equipment was needed during the busy season, it first had to be assembled and painted before use.

There are hundreds of pieces in a single beehive. Each super has four sides, dozens of nails, and frame supports that must be assembled. Each frame has six pieces, more than a dozen nails of different sizes, a sheet of beeswax foundation, and wire to hold it all together. There are eight or ten frames in a box, and each beehive has four—maybe as many as eight—boxes.

Today's manufacturers use plastic and wooden beehive parts and assemble everything at the factory.

An experienced assembler who has all the necessary tools can put together a four-box hive, including frames, in about four hours. And, once assembled, it needs two coats of paint to protect it from the weather.

A first-timer, with most of the tools, could do the same thing in, maybe, two days. For someone with only a passing interest in woodworking and with minimal tools, the task could take a week. If you're not exactly sure where your hammer is—*right now*—you are probably one of these people.

There were, essentially, no options 100 years ago. If you wanted to keep bees, you had to spend the time putting all those pieces together. For some, this is the best part of having bees. In fact, some beekeepers revel in starting from scratch and making their own equipment. They own, it must be noted, workshops that rival the one you see on a certain public television woodworking show.

Don't get me wrong. There can be an untapped, self-fulfilling satisfaction in working with freshly cut wood, fragrant beeswax sheets, and the pleasant hours spent alone, or with a partner, in the aromatic assembly of hives. You may discover this joy in the journey—and while those pieces wait to be assembled, it seems like time stands still.

But these days, the journey isn't the goal for many people. It's having bees in the garden. This is where technology, labor, and the eternal press of time come together. There's now a full range of assembly choices, ranging from the traditional build-it-yourself kits to ready-to-paint, fully assembled hives. If you choose the traditional route and build your own beekeeping equipment, be forewarned that the assembly instructions that accompany these kits are often woefully inadequate. But then, so are the typical instructions for assembling a propane gas grill. The comparison is appropriate, and choosing a preassembled unit over a kit—because self-assembly is a pain, is often poorly instructed, and requires a variety of tools—is a popular choice. Either way, you'll need this book to explain it all.

You will find that some pieces of beekeeping equipment always come preassembled, but completely preassembled beehives are relatively new and have made beekeeping not only more enjoyable but more practical for beginners and seasoned professionals alike. And, it's not only the time savings that has led to this evolution.

The simple realities of having neither the right tools readily available nor a practical place to use them are a couple of reasons why assembling equipment has become so difficult. The garage—

filled with cars, bikes, lawn mowers, garden tools, and the other stuff of life—generally does not have a built-in workshop. If used as a hive-assembly area, especially over a period of time, something has to give. When the task is complete, the "stuff of life" needs to be put back ... somewhere. Basements are just as inconvenient to use as workshops, and few urban or suburban dwellers, which most of us are, have a barn or shop building out back. Basically, dedicating a space large enough to build what you'll need, and having all the woodworking tools to accomplish the task has become problematic and distracts from what hobbyists really want to do in the first place—keep bees in the garden.

So once you've wisely decided to use preassembled equipment, you'll find there are still more choices to make and questions to answer. For instance, what are your physical limitations? The common brood box—called a *deep* because of its height—when full of honey and bees weighs nearly 100 pounds (45 kg). This may be all right for weightlifters and sturdy teenagers to lift, but smaller boxes, called *mediums*, weigh in at about 60 pounds (27 kg) and are a better alternative for the average-strength beekeeper. Using the traditional setup, a typical beehive has two deep boxes and three, maybe as many as five, of the medium boxes. That's a lot of pieces to put together and a lot of lifting when they are full.

Let's simplify this. There are boxes available that hold only eight frames, instead of the ten in the traditional boxes. Better yet, they are available only in the medium size. Best, they come assembled. One of these, when full, weighs in at only 30 pounds (14 kg) or less. Weightlifters need not apply.

Tradition, then, has dictated that beekeeping equipment comes to you in hundreds of pieces, with inadequate assembly instructions that require a variety of tools to assemble, plus the space and time required to put them together.

Wait, I'm not done yet—there's more to this tradition. I've named it the Zucchini Complex. Here's how it works: For springs eternal, gardeners have looked at their large, empty, fertile backyard spaces and imagined them overflowing with the perfect season's harvest. They see great, green, growing mounds of peppers and tomatoes, cucumbers, melons, radishes and beans, okra and greens, summer squash and winter squash, and carrots and corn. And every year, they plan and plot, order seeds and sets, and more.

I grew up in west-central Wisconsin, not far from Minneapolis. Though our neighbor's heritages were mixed, the common

ground among them was dairy farming to earn a living and gardening to feed the family. Because of the blended European backgrounds, rutabagas and Roma tomatoes were grown side by side. But zucchinis were everywhere. They were fast-growing, pest free, and in the spring, while still in the seed pack, nearly invisible.

You know the "August story." Every innocent zucchini seed, planted with love and care in May became a volcano of great, green fruit in August. If gardeners went away for a weekend, they grew to baseball-bat size. Three days of rain yielded three bushels of zucchini, with three zucchini to the bushel.

We couldn't give them away because everybody already had too many. Mysterious mountains of zucchini appeared overnight on the side roads just outside of town. All this sprang from an innocent handful of seeds planted in May. That's the Zucchini Complex. Unfortunately, this complex also applies to beehives.

If you use traditional equipment, good management, and have even average weather, you'll end up with around 100 pounds (45 kg) or so of that wonderful liquid gold—honey—that your bees produce from each one of your hives. One hundred pounds—per hive. To look at it another way, that's nearly two 5-gallon (19 l) pails.

But this is more than tradition. It is the absolute goal of beekeepers everywhere. Those beekeepers, that is, who are intent on enjoyable beekeeping, sustained growth, a fair amount of labor and lifting, and profitable honey production. But that's not *our* goal, not yet anyway.

The solution, of course, is obvious, whether for zucchini or honey. If your goal is *not* to produce record-breaking crops but, rather to learn the ropes, enjoy the process, and not be overwhelmed, then the best way to begin is to start with one, or better, a couple of hives in size eight rather than ten, and manage them so that monstrous honey crops don't overwhelm you with work and storage problems.

Promoting the concept of having bees that don't require hours and hours of work and that produce the size and type of crop that we can manage is the goal of this book.

A long-time friend of mine who is an experienced beekeeper, teacher of the craft, and keen observer of the people who keep bees, once said that people start keeping bees because of the bees, but they quit because of the honey. I'm going to make sure that doesn't happen to you.

CHAPTER 1 ✦ In the Beginning ✦

A beehive should be visually screened from your neighbors, the street, and, perhaps, even your family. The site should have some shade, lots of room to work, and a low-maintenance landscape. Notice that the white hive seen here is highly visible.

Keeping bees is an adventure, an avocation, and an investment, much like preparing for a garden. Considering the amount of sun, shade, and water drainage your yard provides, you must plan where your garden will be and how to prepare the soil. You must make an educated decision about what you can grow and what kind of care your crops will need. You will also need to be aware of harvest dates, and to avoid letting a lot of work go to waste, you'll need a plan for how to preserve the bounty. And, finally, you need to plan what needs to be done to put the land to rest for the off-season. The same planning process applies to beekeeping.

First Steps ✦ Where Will You Put Your Hive?

Your first step is to order as many beekeeping catalogs as you can find. They're free, and they contain a wealth of information. There are also a few magazines dedicated to beekeeping, and a free copy can be had for the asking. (See Resources on page 160.) Look particularly at those companies that offer preassembled products.

Next, read this book. Its chapters explore the biology, equipment, management, and seasonally organized activities of bees and beekeeping. It is important to become familiar with the seasonal routine of beekeeping. It is remarkably similar to scheduling your garden, but the specifics differ and need attention to master them.

Providing Water

Providing fresh water for bees is mandatory. A summer colony needs at least a quart (liter) of water every day, and even more when it's very warm. Making sure that water is continuously available in your yard will make your bees' lives easier, and it helps ensure that they do not wander where they are not welcome in search of water.

Water is as necessary to your bees as it is to your pets and to you. Whatever watering technique you choose for your bees, the goal is to provide a continuous supply of fresh water. This means while you are on vacation for a couple of weeks, when you get busy and forget to check, and especially when it's really, really hot—bees always need water. It is not likely that they will die, as insects are very industrious, but worse, they will leave your yard to find water elsewhere. Suddenly, lots of bees may appear in a child's swimming pool next door or in your neighbor's birdbaths. Outdoor pet water dishes become favorite watering stations for bees on the hunt for water. Bees need water in the hive to help keep the colony cool on warm days, to dilute honey before they feed it to their young, and to liquefy honey that has crystallized in the comb. To make water accessible to bees, try the following:

> Float pieces of cork or small pieces of wood in pails of fresh water for the bees to rest on while drinking.
> Install a small pool or water garden, or have birdbaths that fill automatically when the water runs low.
> Set outside faucets to drip slowly (great for urban beekeepers), or hook up automatic pet or livestock waterers.

Join the Club

There is comfort in company, and even if you're not the type of person who joins clubs or group activities, I encourage you to find a local beekeeping club and touch base with them. (See Resources on page 160.) Most clubs offer beginners' classes, but the greatest value comes from meeting other beekeepers. You'll find that most of them think like you; they are hobby—rather than professional—beekeepers who have limited time but a great interest in honey bees and beekeeping. Many will be able to offer different perspectives and values regarding beekeeping.

Keeping Bees in Your Neighborhood

You probably know of neighborhoods that don't welcome weedy lawns or loose dogs or cats. Some areas also have restrictions on beekeeping. You need to find out about the ordinances of your city or town, because local zoning may limit your ability to keep bees. There are seldom regulations that do not allow *any* beehives on a suburban lot, but there are often specific, restrictive guidelines for managing those that you can have. However, some places strictly forbid having bees. Dig below the surface to find out everything you can before beginning.

It is also important that you investigate your neighbors' take on your new hobby. It may be completely legal to have bees on your property, but if your neighbors don't tolerate your interest, you'll have to make some compromises. People's reactions to bees and beekeeping can be unpredictable. A few will be enthusiastic, most won't care one way or the other, and a few may have an extremely negative opinion of insects that sting and swarm. It's that last group you need to work with. If you are determined to keep bees, a little knowledge will go a long way, and there are some things you can do to allay a reluctant neighbor's concerns.

Often, the cause of a negative reaction from a neighbor is because of someone in the family being allergic to bee stings. Without being confrontational, you should find out if that person is really allergic to bees. Often people lump all flying insects together and yellow jackets or wasps may be the problem, while honey bees are actually not. It is true that a small percent of the population does have a life-threatening allergic reaction to an insect sting (just as some have serious allergic reactions to peanuts or shellfish, for example). Most, however, have a temporary, normal reaction. Bee-sting symptoms include slight swelling at the site of the sting and a day or two of itching and redness. This is the typical response to a honey bee sting and should be expected. However, this book is not a medical text. You, your family, and your cautious neighbors should be very certain about allergic reactions to honey bee stings before you introduce a hive. Do not be alarmed, but do be careful.

Positioning Your Hives

Once you have considered everyone else's comfort level, it is a good idea to consider the comfort and happiness of your bees. Every family pet, including bees, needs a place that's protected from the afternoon sun and sudden showers and provides access to ample fresh water. Bees should be given the same consideration. Place colonies where they'll have some protection from the hot afternoon sun. A bit of shade is good for both the bees and for you. All day sun is alright, but a bit of afternoon shade also affords comfort for the beekeeper when working on a hot summer day

Hive Stands

A hive sitting on damp ground will always be damp inside, creating an unhealthy environment for bees. To keep your hives dry on the inside, set them on an above-ground platform, called a *hive stand*. Before you choose a hive stand, consider that the closer your hive is to the ground, the more you'll have to bend

Cinder blocks are inexpensive, durable, and large enough to support your hives. Set cinder blocks directly on the ground, then place stout 2" × 4" (5 × 10 cm) or 2" × 6" (5 × 15 cm) boards, as shown, between the blocks and the hive. By the end of the season, this durable hive stand may be holding several hundred pounds of hive and honey.

and lift, and the more time you'll spend stooped over or on your knees as you work. This is an uncomfortable way to work, and it makes a good argument for using a raised hive stand. A 2' to 3' (0.6 to 0.9 m)-high stand, strong enough to support at least 200 pounds (91 kg) is ideal. You can build a simple stand using cement blocks and stout lumber. Another option is to make a stand completely from heavy lumber or railroad ties.

Build your hive stands large enough to set equipment and gear on while you work with the hives. If your colonies are placed at the recommended 2' to 3' (0.6 to 0.9 m) above ground, you'll need a spot to rest tools and equipment on during inspection. If your hive stand is small, you will be forced to set the equipment on the ground. When finished with your work, you will have to bend over and lift parts all the way to the top of the hive to replace them. You will be better off creating an additional stand or additional room on one stand on which to set equipment. There is an old saying that is absolutely true: All beekeepers have bad backs, or will have. It is worth the extra planning to avoid the pain.

Making Space

While putting everything together in your backyard—installing the visibility screens and your hive stands all at the right distance from your property line, and perhaps next to a building—you want to be careful not to box yourself in. Plan to have enough

Tip: Keeping Hives Above the Fray
KEEPING YOUR HIVES HIGH AND DRY OFFERS PROTECTION FROM SKUNKS. THESE FRAGRANT VISITORS ARE NOTORIOUS FOR EATING BEES.

elbow room to allow you to move around the circumference of your colonies. This is especially true for the back of your colonies, where you will spend most of your time when working with the bees.

Grass and/or weeds are landscape elements that need to be taken into consideration as well. Left to grow, weeds can block the hive entrance, reducing ventilation and increasing the work of forager bees flying in and out of the hive. It is a good idea to cover a generous area around your hive stand with patio pavers, bark mulch, or another kind of weed barrier. Gravel or larger stones will work if you place a layer of plastic on the ground before installation. Even a patch of old carpeting will keep the weeds down and keep your feet from getting muddy in the spring or after a few days of rain. And a "grow-free" area cuts down on the chance of grass clippings being blown into the front door of a colony.

Placement

The opening of the hive can face any direction that's convenient for the traffic flow of people and bees. It's not critical which way it faces; just remember that your family uses your yard, and being able to keep your bees in check is important to everybody. Finding the best location for your hive will undoubtedly be a compromise between what you, your neighbors, your family, and your community consider important. Once you have decided on the best place for your hives, you have to consider the hive itself.

Good Fences Make Good Neighbors

Being a good neighbor includes doing as much as you can to reduce honey bee/neighbor interactions. Even if you have perfect neighbors, cautious management is an important part of your beekeeping activities and management plans. Here are some important considerations.

> Bees establish flight patterns when leaving and returning to their hive. You can manipulate that pattern so that when the bees leave the hive, they will fly high into the air and away, and then return at a high altitude, dropping directly down to the hive. There are several techniques for developing this flight pattern that will also enhance your landscape. Siting a fence, tall annual or perennial plants, a hedge of evergreens, or a building near the hives will help direct bees up and away from the hive. That same screen will also visually screen your hives from outsiders.

> Neutral-colored hives are less visible than stark white ones. Choose paint colors such as gray, brown, or military green, or use natural-looking wood preservatives. Any paint or stain formulation is safe for bees if you apply it to the exterior of the hive and allow it to dry before installing bees.

> Keep your colonies as far from your property line as possible, within any zoning setback restrictions.

> Avoid overpopulation. You should not have more than a couple of colonies on a typical lot of less than an acre.

Bee Space

When honey bees move into a natural cavity, such as a hollow tree, they construct their nest by instinct, carefully producing the familiar beeswax combs that hang from the top of the cavity and attach to the sides for support, extending nearly to the floor of the cavity.

To keep that spatial comfort zone called *bee space*, they leave just enough room between their combs so they can move from one comb to another, store honey, take care of their young, and have some place to rest when they aren't working or flying outside the hive. This space is not random. Measured, it is not less than ¼" (.6 cm) and not more than ⅜" (1 cm). This distance does not vary between a natural cavity and a manmade beehive, and honey bees are unforgiving if presented with larger or smaller spaces. If there is a space in your hive larger than ⅜" (1 cm), the bees will fill it with beeswax comb in which to raise brood or store honey. If the space is smaller, they seal it with propolis. They do this to ensure there is no room in the nest for other creatures.

There are a couple of exceptions when it comes to comb building of which you should be aware. Bees won't fill the space between the bottom board and the frames in the lowest box in a hive. They leave this space open to accommodate ventilation; the fresh air coming in the front door could not circulate through the hive if comb came all the way to the floor. Generally, honey bees also won't fill the space between the inner and outer cover. This rule is broken only when there is a lot of available food and not enough room in the hive to store it.

Pictured is an eight-frame hive, right out of the box. It has three medium supers, a flat roof, a screened bottom board, and a mouse guard in place.

Equipment & Tools of the Trade

Hives

We've already looked at the basics of the beehives you'll have. Seriously consider using preassembled, medium-depth, eight-frame boxes and appropriate frames. Amazingly, there are no standardized dimensions in the beekeeping industry. The dimensions of hives are not quite the same from one manufacturer to another. As a result, the parts of your hive may not quite fit together if you mix parts from different manufacturers. If your boxes don't quite match, your bees will adjust. But their best efforts to hold the hive together in ill-fitting boxes work against your best efforts to take it all apart when checking on your bees. Sticky, runny, dripping honey from a broken burr comb (a free-form honeycomb built to bridge a gap between hive parts) makes a mess and will cause a great deal of excitement for your bees. Bees will weld ill-fitting boxes together (with a glue called propolis, which they make from plant resins) so that the boxes become inseparable from adjacent boxes. The lesson: In the beginning, choose a supply company carefully and stick with it. Your first consideration should not be cost but ease and comfort for you and your bees.

To get a start in beekeeping, you'll need at least three eight-frame, medium-depth boxes for each colony. You'll soon need a couple more, but we'll explore those options later. Frames hang inside each box on a specially cut ledge, called a *rabbet*. Frames keep the combs organized inside your hive and allow you to easily and safely inspect your bees.

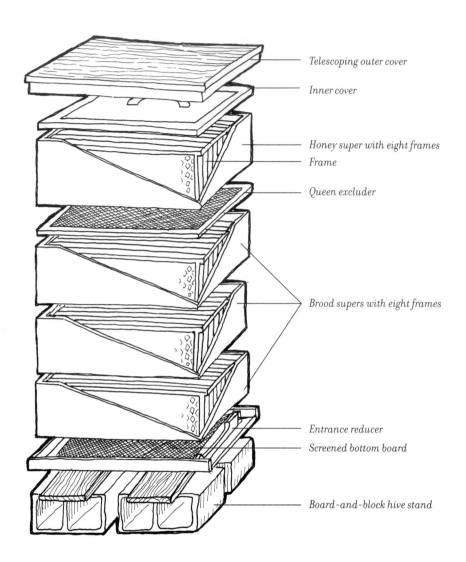

Telescoping outer cover

Inner cover

Honey super with eight frames

Frame

Queen excluder

Brood supers with eight frames

Entrance reducer

Screened bottom board

Board-and-block hive stand

This diagram shows all the parts of a typical four-box hive, which has eight frames to a box. Supporting the hive is a sturdy hive stand composed of concrete blocks topped with boards, which distribute the weight of the hive. A hive can top out at 200 pounds (91 kg) by season's end.

Bee space, shown here, is the space, or gap, between the top bars of frames in a hive. It is also the distance between the top of the frames and the top edge of the box. Bee space allows bees to walk about the hive. If the space is too large, the bees fill the space with honeycomb. If too small, they fill the space with propolis.

An obvious bee-space violation is pictured here. The bees had enough room to build comb and raise brood in the space between the top of this top bar and the bottom of the bottom bar above it.

The History of Frames

Frames were put into practice in the mid-1800s and changed beekeeping from a destructive process to one of benign inspection and harvest. The Reverend L. L. Langstroth, of Philadelphia, is credited with the discovery, but he based his design on the research and work of many others.

All boxes are similar, but there are minor design differences between manufacturers. The primary difference is how deep the rabbet is cut. Deep cuts allow frames to hang lower in the box than shallow cuts. When a box of frames is placed on top of another box of frames, there should be a sufficient "bee space" ⅜" (1 cm) between the two boxes. (See page 19) If a frame hangs too low or too high when the boxes are combined, there will be too much or too little bee space between them. Either scenario makes manipulating the frames, the boxes, and your bees difficult. To avoid this situation, stick with a single supplier when adding or replacing equipment.

Frames

Beehive frames comprise narrow wooden or plastic rectangles that surround the comb. The outside provides support and maintains the rectangular shape of the frame. Bees build their honeycombs within the frame.

Brand-new frames start with the outside support and a sheet of what is called *foundation* within the frame. Foundation is a sheet that is embossed with the outline of the six-sided beeswax cells that bees build. One kind of foundation is made of pure beeswax, complete with the embossed cell outlines. These sheets are fragile and usually have vertical wires embedded in them for support. When assembling traditional frames with beeswax foundations, you frequently need to add horizontal wires for additional support. An alternative foundation is a sheet of plastic that is embossed like the beeswax sheets. These do not need supporting wires. There are also frames made completely of plastic. The outside support and the foundation inside are a single piece of extruded plastic.

You can purchase unassembled wooden frames that come with beeswax or plastic foundation sheets. Assembled wooden frames are also available and come with plastic foundation. If the frames

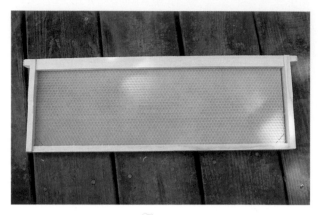

This frame fits medium supers. It has a wooden exterior support that frames an embossed beeswax-covered sheet of plastic foundation in the center.

Make certain you have a screened bottom board. The one seen here has a removable tray beneath the screen to allow monitoring for varroa mites. The tray can be either front or rear loading. Rear is best.

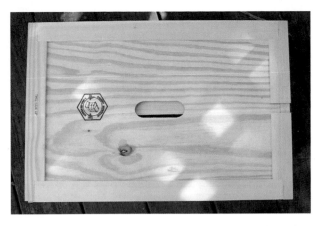

🐝 *An inner cover sits on top of the uppermost super but beneath the outer cover. It has an oblong hole that allows ventilation, feeding, and escape. It has a flat side and a recessed side. The notch provides an upper entrance when needed.*

🐝 *An assembled hive with the individual parts offset, showing from bottom to top: bottom board; three supers and frames; an inner cover; and a flat, metal-sheathed outer cover. An entrance reducer, which doubles as a mouse guard, rests on the outer cover.*

you purchase have plastic foundation, make certain they come with a thin layer of beeswax coating applied to both sides.

The suppliers who sell preassembled boxes also sell preassembled frames that fit in the boxes so that proper bee space is preserved. These are a good match and make setting up a hive much easier.

Bottom Boards

You'll need a floor for your hive. Although several styles are available, consider using ventilated bottom boards. Instead of having a solid wood bottom, these have a screen covering the bottom. Screened bottom boards are good for several reasons: The open bottom provides ample ventilation from top to bottom inside the hive, removing excess moist air and aiding the colony in temperature regulation, and an open floor allows the colony's debris to fall out rather than accumulate on the floor inside. You should, however, make sure there is some kind of a solid slide-in temporary floor.

Inner Covers

Set on top of the uppermost box is an inner cover. If the outer cover is the roof, the inner cover is the ceiling of your hive. It provides a buffer from the hot hive top in the summer and helps regulate air flow. There is an oblong hole in the center of the inner cover. Almost all inner covers are sold preassembled. They are often made from of a sheet of masonite or a patterned paneling. These work but not well enough. They tend to sag as they age. However, some inner covers are made of several thin boards in a frame, which won't sag as they age. Find a source for the latter, as they are worth the search.

Additional items you'll need include a pail-type feeder, an entrance reducer, a bee brush, and a fume board. Each item is explained later in the book, according to when they are used during the season.

Bee suits come in two styles: jackets and full coverage. Full-coverage suits protect your clothes from wax, honey, and propolis, and they also keep the bees out of places where you don't want them. Full suits are good for heavy-duty work. Jackets provide less protection than full-coverage suits. Plastic-covered gloves are commonly used, fairly durable, and moderately good for fine motor skills.

Tip: Elastic Straps
Long-legged bee suits with cuffs have elastic or closing snaps that make crawling bees a nonissue. But, because a determined honey bee can make it an issue, having these elastic straps and not needing them is wiser than needing them and not having them. Keep a pair in your back pocket.

Personal Gear
Bee Suits

A bee suit is your uniform, your work clothes, what keeps you and your bees at a comfortable distance, and what keeps your clothes clean. To meet the needs of the individual beekeeper, the sophistication and variety of bee suits is first rate. You'll find that white is the most common color, but any light-colored suit is acceptable. Full suits cover you from head to foot but are quite warm in summer weather. An alternative is a bee jacket. These are cooler, but they don't keep your pants clean. The important thing to keep in mind when working with honey bees is that they are very protective of their home. When anything resembling a natural enemy approaches, such as a skunk, bear, or raccoon, they will feel threatened. These enemies have one thing in common—they are dark and fuzzy—so, wearing dark and fuzzy clothes near the hive is not a good idea. Whichever bee suit style you pick, keep it simple to start, and get one with a zipper-attached hood and veil. These offer good visibility, durability, and *no* opportunity for an errant bee to get inside. And because the veil is removable, you can try other head gear later without having to invest in a whole new suit.

When you're examining your colony, bees will land on your suit and your veil, and they'll walk on your hands. This isn't threatening behavior, but initially it can be distracting and a little disconcerting. Wearing gloves can remove that distraction. Most people wear gloves when they start keeping bees, and most quit wearing them after a while. The cardinal rule is to wear what makes you comfortable.

Gloves

You can buy heavy, stiff leather gloves, which are made for commercial beekeepers, but our goal—as hobbyists—is finesse, not hard labor. I recommend buying the thinnest, snuggest, most supple gloves you can find. A common style is made of thin, plastic-coated canvas material. Long cuffs, called gauntlets, are attached. Gauntlets slip over your long sleeves to keep bees from climbing into your sleeves.

Some suppliers sell gloves in exact sizes (not the traditional S, M, L, XL) and these, usually made of thin, soft leather, will fit best. This is especially important for the fingertips. Glove fingers that are too long make you clumsy and awkward, and it's difficult enough to be careful when moving frames. Regular rubber dishwashing gloves work well too, offering excellent dexterity when handling frames.

🐝 *Your smoker is indispensable when working with bees. Shown here is a good-sized smoker. The beekeeper is wearing an attached-hood jacket and thin leather gloves for protection. Boots keep your feet dry and keep bees out of your cuffs.*

When you are ready to give them up, not wearing gloves is the best way to go for most beekeeping activities. But everyone has his or her own schedule for reaching that level of comfort. Eventually, you'll cut the worn-out fingers off an old pair of gloves to increase your dexterity. Then one day, you'll forget to put them on completely and not even notice.

Ankle Protection

Something not often thought about until it's too late is the gap between the tops of your shoes and the bottom of your pants. We seldom think of bees as being on the ground, but when you open a hive to lift out a frame or move boxes, bees will fall out. Most will fly away, but some won't. These are the bees you need to be aware of, because these bees will crawl—especially if the weather is cool or they are young and not used to flying. Sometimes, a lot of them will drop to the ground in a bunch and crawl for a bit before they get their bearings and fly away. This is especially true if they land in grass or weeds, rather than on a smooth, flat surface.

Bees that land on the ground naturally crawl up something. Usually their options are climbing up the hive stand or on your shoes. To avoid the latter, beekeeping suppliers sell elastic straps with hook and loop attachments that are easy to use.

Smokers and Fuel

A smoker is a beekeeper's best friend. A simple device that has changed little during the past 100 years, it is basically a metal can (called a fire chamber) with a hinged, removable, directional nozzle on the top, a grate near the bottom to keep ashes from blocking the air intake from the bellows, and the bellows. Only large and small sized smokers are available, and the large model, no matter who makes it, is the better choice. Stainless steel models last longer than galvanized-metal ones, and a protective shield on the outside of a smoker is there for a good reason. Buy a large, stainless steel model with a shield.

What do you burn in a smoker? Many fuels work well, but some are dangerous. Beekeeping supply companies offer fuel cylinders made of compressed cotton fibers and small pellets of compressed sawdust. Have some of these available at all times, because there will be times when your other fuel is wet or you are out of your regular fuel source.

Many types of fuel are plentiful and free for the taking: Sawdust is one, chipped wood mulch is another, and pine needles are also wonderful. Dry, rotten wood—called punk wood—which is soft enough to crumble in your hands and can be collected during walks in the woods, is ideal. Small pieces of dry wood left over from a building project work well, too, as long as they fit

How Smoke Affects Bees

Thousands of years ago, someone figured out that if he had a large, burning torch with him when he went to rob a wild honey bee nest of its honey, he would have a much easier time of it. The smoke from the fire calmed and quieted the bees somewhat while they were being robbed.

Several things happen when you puff a bit of smoke into your colony. The primary form of communication in a honey bee colony is odor—when you puff in a bit of smoke, it masks odors and effectively shuts down communication, causing, understandably, quite a bit of confusion. The normal order is disrupted, and the chain of command is broken. This organizational breakdown allows a window of opportunity for a beekeeper to open the colony, examine what needs to be examined, do the work that needs to be done, and close up the hive before order is restored.

When the smoke first enters the colony, some bees simply retreat from it. They literally run to the most distant part of the colony to escape the smoke. These are mostly house bees, which are too young to fly. Others head directly for the nearest stored honey and begin eating as fast as they can. It is suggested, by scientists, that this behavior occurs so that in the event that the colony needs to abandon its nest due to a fire, some bees will leave with a full load of supplies needed to sustain them while a new nest is constructed.

Some bees, however, seem not to be affected by smoke. These bees tend to be the guard bees that work on the periphery of the nest, and they are not as influenced by the communications that go on within the hive. Their tendency is to fly about, explore, and attend to the disruption.

However, even guard bees have some level of odor communication, which is disrupted by smoke. Ordinarily, when a honey bee senses danger in a hive, she emits an alarm pheromone. This pheromone has a banana odor that can even be detected by the human nose. When other bees detect the alarm pheromone, their instinct is to investigate the cause of the problem. If the threat is real, some bees may sting, which will release additional alarm pheromone into the mix. Bees will continue this activity until the threat is removed. With smoke present, however, even guard bees, which aren't in the colony and aren't affected by the disruption, will find it difficult to alarm other inhabitants. Although this situation may sound like a melee, it actually leaves the beekeeper to work in relative peace.

Smoking can be overdone, however, and once you have completely confused the colony, the disorienting effect is eventually negated, and, in fact, confused bees will begin flying no matter how much smoke you use.

into the fire chamber. Untreated jute burlap is good fuel, but be careful not to use synthetic burlap. Untreated twine from baled hay or straw also can be used, but beware—both burlap and twine are often treated with fungicides or other antirot chemicals so they don't disintegrate in wet weather. Make sure you are burning untreated materials. Don't use petroleum-based fire starters or gasoline. Bees are sensitive to chemicals, and the fumes from treated materials would kill your bees and probably cause flare-ups and other fire-safety problems in your smoker.

Hive Tool

Beekeeping supply catalogs offer several styles of hive tools. The most utilitarian is the one that looks like a paint scraper with one end curved and the other end being broad, flat, and sharpened, and the standard 10" (25 cm) hive tool will provide the most leverage. Hive tools are inexpensive and, interestingly, easily lost. I recommend starting out with two. Other styles are designed for specialized tasks in a beehive, and you'll see their advantages when you've had some experience with the standard hive tool.

🐝 *The curved end of the most commonly used type of hive tool is used as a scraper, lever, and hook. The long, flat end is a scraper used to remove burr comb and propolis and to ease frames out of the supers.*

Don't be Fooled

You will see tools that look like a regular hive tool in most hardware stores, but there is a significant difference between these and the tools sold by beekeeping supply companies. These hardware-store paint scrapers are not tempered and will break when used to pry apart your supers or to remove frames. Hive tools are a hard-working part of your gear, so don't take chances with a tool not intended for this use.

The Bees
Packages and Nucs, Colonies and Swarms

Beekeeping equipment is famously specialized. It is certain that your local hardware store, farm store, or discount home center won't have anything you need. This is especially true when shopping for the actual honey bees. Here's where being part of a local beekeeping club or having taken a beginner's beekeeping class is to your advantage. Knowing somebody who knows somebody who has whatever it is you'll need, whenever you need it, is the key to successful beekeeping.

There are several ways to obtain bees for your hives—two easy ways and a couple of other ways that are far more exciting and immediate. You can buy what is called a *package* of bees, which is simply a screened box containing honey bees and a queen, which will be shipped from a beekeeper who grows bees especially for this purpose. You will transfer these bees into your own hives, get them started, and keep them going as a colony. Or you can buy a small starter colony, referred to as a *nuc* (short for nucleus colony), that you install in your own hive.

Another option is to purchase a full-sized, ready-to-go colony of bees from another beekeeper. The advantage to this is that you minimize the risks of starting a small, somewhat vulnerable nuc, but the potential disadvantage is that you start out at full speed, without the breaking-in period that most beginning beekeepers need to establish their own comfort level with the craft.

Catching a swarm of bees is how some beekeepers get their start. This entails finding, capturing, bringing home, and hiving a swarm of honey bees. This activity is as exciting as beekeeping gets. (See Catching Swarms on page 109.)

It's All in the Preparation

So far, we've looked at the tools you'll need to get started, reviewed the pieces and parts of hives, and planned where your hives will be located when they are up and running. We've also looked at your work gear—the protective suits and gloves, smokers, and hive tools—and where your bees will come from and how large the starter colony should be.

The old motto of always being prepared goes without saying, but I'll say it anyway. Start your preparations early: make sure you have everything you need; make all the helpful contacts you can; read the beekeeping catalogs, journals, and books, especially this book; and if at all possible, find a local club and take a starter course in beekeeping. And make certain that your neighbors and your family support your beekeeping aspirations. Now the adventure begins.

Buying Packaged Bees

Early spring arrives two or three months earlier in warmer regions than in more moderate and cooler regions, no matter where you live on the globe. People who live in warm regions and produce bees to sell start raising bees very early in the year, so they have them ready to sell when spring arrives in cooler areas.

In order to do this, they remove some bees from their colonies every three weeks. They open a colony, find and remove the queen, and shake excess bees into a package (a screened cage) made especially for shipping live bees. The most commonly sold amount is a 3-pound (1.5 kg) package of bees, but 2- and 4-pound (1 and 2 kg) packages are also available. A 3-pound (1.5 kg) package is about the right amount for one eight- or ten-frame hive. There are about 3,500 live bees to a pound, so your 3-pound (1.5 kg) package will contain about 10,000 bees.

A can of sugar syrup supplies the bees with food for several days. A queen, snug in her own protective cage, is kept separate from the bees in the package because it takes a few days for the packaged bees to become acquainted with her. This complete package is shipped directly to a customer or a local supplier. Chapter 3 discusses how to get the bees and the queen from the package into your hive.

If you're lucky, somebody in your local club will have truckloads of packaged bees shipped directly to his or her place of business to sell in the spring. Check local suppliers before ordering, because it is best to buy locally. Find out what they are selling (the size of the package or nuc), the cost, the day the packages will be available (generally there is only a small window of opportunity—a weekend is common), and what choices for types of bees or queens you will have. Find out, too, where the suppliers are getting their bees and the queens and how long it will take for bees to be shipped. When it comes to price, the saying "you get what you pay for" is mostly true. If you live within a few hundred miles of primary suppliers, you may be able to buy directly, or have bees shipped to you through the mail. However, bees can be shipped only limited distances before the stresses of travel take their toll.

Buying Nucs

Often, the same people who sell bee packages will have nucs for sale. A nuc has three or four frames instead of eight, and contains a small, already-established colony, complete with a laying queen. Nucs provide a head start on the season and take some of the risk out of trying to start a colony on your own.

Nucs usually cost more than bee packages. However, a prob-

lem to be aware of with nucs is the compatibility of equipment. Before you decide to buy a nuc from a supplier, double-check the size of the frames in the nucs. Often, nucs come in deep frames, not the medium size this book recommends. If you ask ahead of time, your supplier can probably make a nuc the size you specify. Many more suppliers are starting to make nucs in a medium size due to the popularity of medium-sized equipment. But don't make that assumption—ask before you buy.

Buying a Full-Sized Colony

Another way to get started with bees is to buy a full-sized colony from another beekeeper. This approach makes you an instant beekeeper, but it also gives you all of the responsibilities that go along with being a beekeeper. You should consider a few things before taking this step. First, in the spring, full-sized colonies will need to be managed for swarm control and monitored for pests and diseases, and will have a large population to deal with. There's no break-in period when you go down this road.

One other factor to consider when buying a full-sized colony is that it belonged to someone else. Like buying anything used, you should have another, more experienced beekeeper or your local apiary inspector evaluate the colony for health and equipment quality before buying.

Types of Bees

All honey bees have a common ancestor, but their natural or man-assisted migrations have allowed for the development of species, or breeds, with adaptive traits. Honey bees now exist in all parts of the world except the two polar regions. Breeds have adapted to survive in deserts; during long, frigid winters; through rainy and dry seasons; and in weather conditions between these extremes. The natural selection process has resulted in honey bees that are very skilled at living in cavities similar to traditional manmade hives, gathering and storing provisions to last during winter when pollen and flower nectar is scarce, and choosing to swarm early in the food-rich spring, increasing their probability of establishing a new nest, storing food, and surviving future winters.

More than 20 breeds of bees have been identified, and many of these have been tested by beekeepers for their ability to live in manmade hives, as well as their adaptability to the moderate climates of the world. Many species have been abandoned by beekeepers because they possess undesirable traits, such as excessive swarming, poor food-storage traits, or extreme nest protection.

🐝 *Italian bees are generally yellow with brown or black stripes. Drones and queens have large, golden abdomens.*

Italians (Apis mellifera ligustica)

Italians are by far the most common honey bee raised in the world. Having evolved on the moderate to semitropical Italian peninsula, Italian bees adapted to long summers and relatively mild winters. They begin their season's brood rearing in late winter and continue producing brood until the beginning of winter or later. Italians never really stop producing young, but they do slow down during the shortest days of the year.

Beekeepers living in southern climates are faced with few management problems. There are nectar and pollen plants available during almost all of the bee's active months. But Italian bees kept in moderate and cool regions are challenged by a shorter growing season to make and store enough food to last through the long winters.

Package producers prefer Italian bees because they can start the rearing process early and raise lots of bees to sell. Beekeepers who pollinate crops for a living also like this trait because they can produce populous colonies in time to pollinate early-season crops. And Italians produce and store lots of honey when there is ample forage and good flying weather.

Italians are also attractive to beekeepers because they are not markedly protective of their hive. Italians are quiet on the comb when you remove and examine frames, they do not swarm excessively, and they do not produce great amounts of propolis.

Italians are yellow and dark brown or black in color and have distinct yellow and dark brown or black stripes on their abdomens. The drones are mostly gold, with large golden abdomens lacking stripes. The queens are easily identified because they have a very large, orange-gold abdomen that is strikingly different from all the other bees in the colony.

◄◄◉►

Carniolan honey bees are dark with brownish to dark gray stripes. Queens and drones have nearly black abdomens.

Carniolans (Apis mellifera carnica)

Carniolan honey bees developed in the northern part of south-eastern Europe in the area of the Carniolan Alps, including parts of Austria, Slovenia, and areas north and east of that region. The mountainous terrain and somewhat unpredictable environment prepared these bees to survive cold winters and to react to quickly changing weather and seasons. As a result, they react quickly when favorable weather arrives in the spring, increasing their population rapidly and swarming early to take advantage of a short season. During the summer, they take advantage of the abundant food, but if drought or other unfavorable conditions arise they can slow their activity just as rapidly. When fall approaches, they slow their activity even more, and during the winter they survive with a small population and consume significantly less food than they do during the growing season.

Carniolans, unlike Italians, are dark in color. The workers are dark gray to black, with gray stripes on the abdomen. The queens are all black, and compared side by side, not as large as Italian queens. Drones are large and have all-black abdomens.

These are the gentlest of all the honey bees. They are quiet on the comb when the beekeeper examines frames, and they tolerate typical beekeeper management duties. They also use propolis sparingly and tend to be a bit more forgiving in situations where burr comb would normally be used.

🐝 *Caucasians are dark gray to black with lighter gray stripes on the abdomen. Queens and drones have dark gray to black abdomens.*

Caucasians (Apis mellifera caucasica)

Less commonly used are Caucasian honey bees, developed in the Caucasus Mountains of Eastern Europe. They reproduce very slowly in the spring and react well to available resources during the summer. Like Carniolans, they respond to winter by reducing their population and using honey stores sparingly. But, because they build slowly, they swarm later in the spring than either their Italian or Carniolan cousins.

Caucasians are extremely gentle to work and are quiet on the comb when being examined. However, they tend to be susceptible to diseases, especially nosema. They also use propolis in every place you can imagine, which makes working your hive extremely difficult.

Caucasian workers are dark gray, with light gray stripes on their abdomens and sometimes brown spots. Queens and drones are dark, like Carniolans.

Other Bee Varieties

Other types of honey bees are available. A commonly available hybrid, Buckfast, is a cross of several breeds. Other varieties reflect selections that have been made within a breed. These selections are made primarily so the resulting bees are easier to manage, produce more honey, or adapt to specific locations. Before you purchase any type of bee, do some research. Check with local beekeepers to see what kinds of bees they buy and how successful those types have been in your area. Look for bees that carry some pest and disease resistance, are hardy enough to winter where you live, and come from a reputable producer. There will certainly be some trial and error before you decide what type of bee you are most comfortable with and which performs best in your location.

One last thing—the descriptions given in this section are, for the most part, the ideal. Carniolans are supposed to be black, but more often than not you will get a Carniolan queen that produces workers with some yellow on them. This is because the producer let some crossbreeding occur. However, the predominant traits of the variety will almost always stand out.

IN THE BEGINNING

29

Checklist of Equipment

Equipment for Each Hive

> Screened bottom board

> At least three assembled,
  medium-depth brood chambers

> At least two additional medium-depth supers
  for honey, which may be assembled regular honey
  supers, complete with assembled frames

> Queen excluder

> Mouse guards for the front door of the hive

> Inner cover

> Cover (Both styles, peaked or flat,
  work well; peaked are decorative.)

> Bee suit (with attached veil), gloves, ankle straps

> Hive tools—at least two

> Smoker and smoker fuel

> Hive stands

> Hive-top sugar-syrup feeder pail

> Books, magazines, and other beekeeping
  information

> Honey bees and a queen

Propolis

Honey bees use propolis, a glue they make from plant resins, for a variety of purposes. Propolis is a sticky, resinous material produced by plants to protect their leaf or flower buds against insects, fungi, and bacteria. Bees scrape small amounts of this material off the buds, without harming the plants, and transfer it to the pollen-carrying parts of their back legs. Enzymes are added by the worker bees to make it workable and to increase its fungicidal and microbiological properties.

Bees use propolis to seal small cracks in the hive, to keep out wind and rain, and to exclude small creatures. They also smooth rough spots by plastering them with propolis, and sometimes reduce the size of the hive entrance by building mounds of propolis around it.

There is a profitable side to propolis. Many people seek it as a raw material for making homemade medicines for scratches, sore throats, and other minor ailments. If harvesting propolis is an option for you, there are companies that purchase it from beekeepers for a fairly good price. Beekeeping supply companies sell propolis traps—plastic mats with small slits in them. These slits are smaller than bee space, and the bees will fill the slits with propolis. When full, the mat is removed from the hive, the propolis harvested, and the mat reused. You can also harvest propolis by scraping it off the hive. Propolis will be a part of your beekeeping life. For honey bees, it is essential. For you, it will be irksome, ordinary, and for some, profitable.

Properties of Propolis
PROPOLIS, LIKE THE PLANTS IT IS COLLECTED FROM,
COMES IN A VARIETY OF COLORS, FROM GRAY TO TAN, GOLD,
BROWN, AND NEARLY BLACK. WHEN WARM, PROPOLIS IS STICKY,
GOOEY, AND DIFFICULT TO WORK WITH. WHEN IT IS COLD,
IT BECOMES BRITTLE.

Removing Unwanted Propolis
IF YOU GET PROPOLIS ON YOUR BEE SUIT (AND YOU *WILL*
GET IT ON YOUR BEE SUIT, SMOKER, HIVE TOOL, AND OTHER
CLOTHES), IT CAN BE DIFFICULT, OR IMPOSSIBLE, TO REMOVE.
BEFORE WASHING, TRY RUBBING IT OFF WITH A RAG DIPPED
INTO SPOT REMOVER OR A PETROLEUM DISTILLATE.
ONCE PROPOLIS-STAINED CLOTHING IS WASHED,
THE PROPOLIS IS SET INTO THE MATERIAL FOR LIFE.

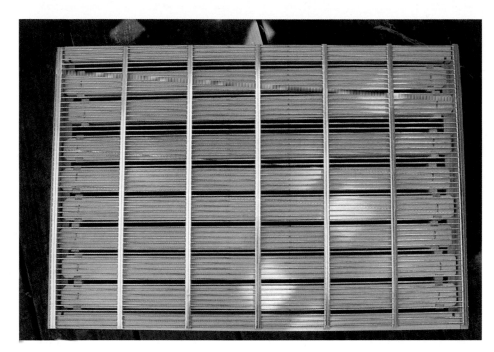

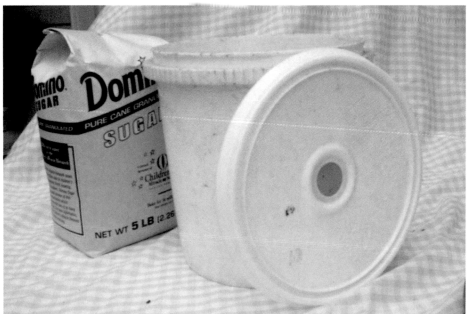

:◀◇▶

Other pieces of equipment you will need include: an eight-frame queen excluder (have at least one for every colony you own, plus an extra)—this one has a metal edge; below, a plastic 1-gallon (3.8 l) sugar syrup feeder pail, with screened opening on top, and sugar.

Time Line for Beginning

The honey bee season follows the growing season, no matter where you live. Once the weather warms in spring and the days are long enough, plants grow and begin to bloom, and the honey bees will begin flying, foraging, and collecting nectar and pollen.

A classic rule of thumb is to plan to have new bee packages arrive a week or so before the dandelions bloom where you live. If you don't know when dandelions bloom, ask a local, experienced beekeeper (or gardener) when the bees ship, and start your plans with that date in mind. You'll need your equipment prepared before that date, and your hive stand and landscape screening set up.

In most years, ordering bees in late winter is good advice. The shortest day of the year is a good time to get an order in for the earlier packages, and you can order them as late as two months after that and still receive packages in the spring. After late spring, package shippers are reluctant to ship for fear the bees will overheat in transit.

CHAPTER 2 →About Bees←

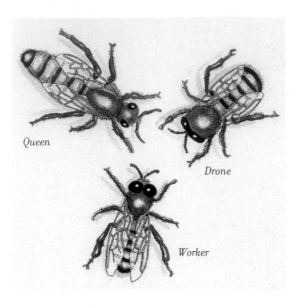

🐝 *Shown here are a queen at top left, a worker at bottom left, and a drone at right*

Queen

Drone

Worker

Overview

Your honey bee colony follows a predictable cycle over the course of an entire season. To successfully manage it, you need a mental picture of what should be happening throughout the year, to help you time your visits, have the right equipment ready, and prevent problems.

You should also be familiar with the individuals within the colony. It is vital to understand the queen, the workers, and the drones, as well as how these individuals interact with each other, how they act and react as a group, and how they respond to their environment. Recognizing any situation that isn't normal is an important step in preventing problems or correcting them when they arise.

Let's start by looking at the individuals in the colony: the queen, the workers, and the drones. We'll explore their development and what each one does during the season. As we do this, we'll also examine the colony as a unit, as well as the bees' environment—including where they live, how seasonal changes affect them, and your interactions with them. In the next chapter, we'll bring all of this together and develop a predictable, seasonal plan that anticipates your activities and how to make beekeeping practical and enjoyable.

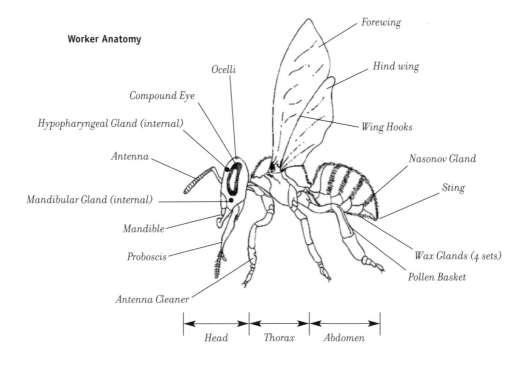

Worker Anatomy

Forewing
Hind wing
Wing Hooks
Ocelli
Compound Eye
Hypopharyngeal Gland (internal)
Antenna
Nasonov Gland
Sting
Mandibular Gland (internal)
Mandible
Proboscis
Wax Glands (4 sets)
Pollen Basket
Antenna Cleaner

Head *Thorax* *Abdomen*

🐝 *This illustration shows the basic body parts of a honey bee.*

The Queen

All bees begin as eggs laid by the queen of their colony. Eggs destined to become queens and those destined to become workers are identical at the egg stage.

A queen bee can fertilize eggs with sperm stored in her spermatheca. As the developing egg passes through her system, sperm is released, and the egg is fertilized just before she places it in the cell. For three days, the egg develops within its shell. On the third day, the eggshell, or chorion, of the egg dissolves, releasing a tiny grublike larva.

Worker "house" bees immediately provide food for these tiny larvae, making a thousand or more visits each day to feed them. This food, for the first two-and-a-half or three days is identical for both workers and queens. It's a rich, nutritious mix, called *royal jelly*, that the house bees produce from protein-rich pollen, carbohydrate-laden honey, and enzymes they produce in special glands. The house bees add the royal jelly to the cells, and one larva floats in each filled cell.

Larvae destined to be royalty see no change in this rich diet and continue to grow and develop. Workers-to-be, however, get a diet change on about day three. Their rations are downgraded in quantity and protein content, which keeps them from developing into queens. This difference, called *progressive provisioning*, allows the royal jelly-fed larvae to fully develop the reproductive organs and the hormone- and pheromone-producing glands necessary

Queen honey bees have long, tapered abdomens and are larger than workers. They vary in color, depending on their breed.

to fulfill their future role as queens. They also mature faster than other bees, completing the egg-to-larva-to-pupa-to-adult cycle in only sixteen days, compared to the twenty-one days required for workers and twenty-four for drones. (See chart below.)

The cells in which queens are raised are different than worker cells. Because of the enriched diet, queen larvae are larger than worker larvae and require more room. Their cells either extend downward, filling the space between two adjacent combs, or hang below a frame. A queen cell is about the size and texture of a peanut shell with an opening at the bottom, making them easy to identify. The smaller worker larvae fit into the horizontal cells of the brood nest.

	EGG	LARVA	PUPA	TOTAL DEVELOPMENT TIME
Queen	3 DAYS	5 ½ DAYS	7 ½ DAYS	16 DAYS
Worker	3 DAYS	6 DAYS	12 DAYS	21 DAYS
Drone	3 DAYS	6 ½ DAYS	14 ½ DAYS	24 DAYS

The diagram illustrates development time in days for queens, workers, and drones.

Because of their large size, queen cells are attached to the bottoms of frames, or fit between frames, and are easily visible.

Queens are produced for a variety of reasons: to replace a queen lost through injury, in preparation for swarming, or to replace a failing, but still present, queen.

When a replacement queen is needed to replace an injured or failing queen, colonies almost never produce only one queen cell; they make as many as they can (if resources are limited) or as many as they want (if resources are ample). Queen cells can range in number from two or three to twenty or more and can be found on both sides of several frames. The process of producing multiple queen cells occurs over two to three days. Therefore, not all of the queen larvae are of the same age. The first queen to emerge destroys as many of the still-developing queens as she can find, eliminating the competition. She does this by chewing through the side of the queen cell and stinging the developing queen pupa inside. Sometimes, two or three queens emerge simultaneously, and they eventually meet and fight to the death, often with help from the workers.

A colony generally tolerates only a single monarch, but on occasion, an older, failing queen and the triumphant daughter can coexist for a time. Sometimes, sister queens who emerge at the same time coexist without fighting. The common thread here is the close relationship and similar chemical cues they produce. In both cases, this is positive for the colony because of the increased egg-laying potential. Eventually, the older queen dies, or is killed by the workers.

For two to three days, the victorious, virginal queen continues to mature, feeding herself or being fed by house bees. Orientation flights near the colony begin after a week or so. The young,

unmated monarch needs to learn the landmarks near the hive so that she can find her way back after a mating flight. Once she is comfortable with navigation, weather permitting, she starts mating. Queens hardly ever mate with the drones from their own colony (inbreeding could cause genetic problems in off-spring). Instead, they take flight, looking for drones from other colonies. Drones and queens gather in places away from their respective colonies, called *drone congregation areas* (DCAs), mating 30' to 300' (9.1 to 91.4 m) in the air above open fields or forest clearings.

A virgin queen emits an alluring come-hither pheromone during this flight, inviting a whole slew of drones to follow. The fastest drone catches her from behind, inspects her with his legs and antennae, and, if he deems her to be a potential mate, inserts his reproductive apparatus. The act stuns and seems to paralyze the drone. His body flips backward, leaving his mating organs still inside the queen. He falls and dies. These organs, called the *mating sign*, are removed by the workers when the queen returns to her hive.

Depending on the number of drones available—and, of course, the weather—a queen may make several mating flights within a few days. She may mate with as many as twenty drones or as few as five or six. Generally, the more the better, because it increases the amount of sperm available and the genetic diversity of the bees this queen will produce during her life.

Occasionally the queen will not mate because of an extended period of bad weather. After five or six days, she will be past mating age, so the colony will raise more queens, if possible.

If not, the colony may go queenless. This situation requires the attention of a beekeeper or the colony will perish.

When the queen's spermatheca is full, her mating days are over, and she begins life as a queen. Prior to mating and during her mating flights, queens are not treated like queens in the hive. They don't begin producing the colony-uniting pheromones until after mating. They do, however, have some chemical control before mating. They can inhibit both further queen-cell production and the development of ovaries in workers, even though the egg-producing organs in her own abdomen—the ovaries and ovarioles—aren't completely matured until her mating begins. The queen appears to grow even larger now as these internal organs expand, but in reality, her abdomen is stretching to accommodate them.

Queens produce several complex pheromones, or distinct odors or perfumes, which are perceived by workers. Many of these chemicals are produced in glands located in the queen's head near the mandibles. According to bee scientists, at least seventeen compounds are produced in this volatile mix, often referred to as queen substance. Several other pheromones are produced by the queen in other glands. As the worker bees feed and groom the queen, they pick up minute amounts of these chemicals. Then, as they go about their other duties, they spread the chemicals throughout the hive, passing along scent cues that inhibit certain behaviors and strengthen the frequency and intensity of others. The most important message relayed by these chemicals is that there is a queen present.

In an unmanaged colony, barring injury or disease, a typical queen will remain productive for several growing seasons. As she ages, her sperm supply is continuously reduced, and her ability to produce all of the necessary pheromones for colony unity diminishes.

There comes a time when the workers in a colony are able to detect that the appropriate pheromone level is no longer sustained. This situation happens for two distinct reasons—overcrowding and supersedure—which in turn elicit two very distinct behaviors in the colony.

Making Wax

By the time a worker is about 12 days old, her wax glands have matured. These four pairs of glands are on the underside of her abdomen. Wax is squeezed out of the glands as a clear liquid. It cools rapidly and turns white. The worker uses her legs to remove the wax, and then manipulates it with her mandibles to build the hive's architecture. Pure beeswax is used to cap filled honey cells or to build new comb for storage. New beeswax is mixed with old beeswax and a bit of propolis, for strength, when covering brood cells and for use in building bridge comb.

When bees build new comb on a sheet of beeswax or beeswax-covered plastic foundation, they are said to be "drawing out" the comb. That is because they use the small amount of beeswax available to start the hexagonal cells, and then add to this foundation new wax that they produce in their wax glands. The result is a frame that looks like this, with all-white wax cells.

🐝 *When a swarm of bees leaves its colony, it fills the air all around the colony. The swarm then heads for a nearby branch or other object on which to settle before moving to its final home.*

Swarming

Overcrowding occurs at the beginning of the growing season, when abundant forage becomes available, the weather is favorable, the population of adult bees is large, and the brood population is rapidly expanding. The colony is crowded with adults, more are on the way, there's little room to expand, and the external environment invites exploration. This situation triggers half the workers in the colony to change from a brood-rearing, foraging mode to one of slowing down production, packing up, and preparing to move.

One result of this situation is that young workers, those able to produce wax, begin constructing the base of large queen cells along the bottoms of the frames in the brood nest, building them so that they hang down from the bottom of the frame. These bases are called *queen cell cups* and you will often notice them along the bottoms of frames. Finding them may indicate the early stages of swarming plans. At the same time, some of the previously foraging bees begin to look for a habitat that would make an acceptable new home.

Because room for expansion in the colony is limited, the queen slows egg-laying behavior, needs less food, and, within three to four days, stops laying eggs completely. A queen who isn't laying eggs loses weight and slims down because her ovarioles shrink. Her newfound slimness allows her to fly—something she hasn't done since her mating flight.

Her changed behavior becomes more intense over the five to six days that the new queen larvae are developing, and those workers that have become aware of the changes are no longer foraging but staying in the hive and gorging on honey—gathering provisions for the move. The final piece falls into place when the first of the queen larvae reaches pupating age and her cell is capped by the workers. This is the final signal for the existing queen to move to a new residence.

If the weather cooperates, the scout bees, which have been searching for a new habitat, and other workers begin racing around in the broodnest area, stirring up the colony. Suddenly the bees—including workers, a few drones, and the reigning queen—leave, pouring out the front door by the thousands.

Staying to keep the home fires burning is the remaining population. They continue to work and live in the colony as if nothing has happened, foraging, ripening and storing nectar and pollen, and tending to the new queen. Meanwhile, the departing swarm fills the air around the colony. It slowly organizes and heads for a nearby location, such as a tree branch or fence post, usually 50 yards (45.7 m) or so from the colony's former home.

Scout bees, those that have investigated possible new homes, join the waiting swarm and perform directional dances on the surface of the swarm to persuade more scout bees to visit the prospective locations. When one site draws more visitors than others, the scouts return and begin the mobilizing activity again. The swarm rises and heads toward its new home. There, new comb is constructed, foragers begin immediately to bring nectar and pollen home, and soon the queen begins to lay eggs. A new colony is complete.

Meanwhile, in the original colony a new queen has emerged and mated and is laying eggs. The colony continues as before, but it has had a break in its egg-laying schedule of about three weeks.

Supersedure

Supersedure, or replacement of the existing queen, occurs not when the colony is in swarm mode, but because either an emergency has occurred or the old queen is failing.

Recognizing a Queenless Colony

Colonies that have lost their queen display some definite behaviors that will cue you in to the situation. These behaviors, however, can occasionally be noted when the colony is queen-right (having a healthy, laying queen). It's not always perfectly clear what's going on, but a close examination usually shows that several queenless behaviors are occurring simultaneously. As soon as an hour after a queen disappears, the lack of her pheromonal presence is pretty well understood by all or most of the bees in the colony. Within a day, sometimes more or less, the bees will undertake emergency supersedure behaviors.

With the queen's pheromone signals vanishing, the colony becomes stressed, leaderless, and without direction. Behaviors include an increase in fanning behavior, seemingly to better distribute what regulating chemicals are remaining in the hive. This fanning is noisy—literally. You will notice the difference immediately when you remove the hive's cover. At the same time, you'll notice an increased defensive level from the guards at the front, and even from those who rise to meet you from the

Tip: When the Queen Escapes

IF THE QUEEN FLIES AWAY WHEN YOU ARE EXAMINING YOUR COLONY (AND YOU ARE FORTUNATE ENOUGH TO SEE HER LEAVE), QUICKLY AND CAREFULLY PUT EVERYTHING BACK TOGETHER, BUT LEAVE THE INNER HIVE COVER HALF OFF AND THE COVER COMPLETELY OFF FOR AN HOUR. OFTEN, THE AROMA OF THE HIVE WILL SERVE AS THE BEACON SHE NEEDS TO HELP HER FIND HER WAY BACK. JUST AS OFTEN, HOWEVER, SHE WON'T RETURN, AND THE COLONY WILL HAVE NO QUEEN.

The first thing bees do in their new colony is produce beeswax combs so that they have a place to store food and raise young.

🐝 *Supersedure queen cells are found on the face of this comb. The queen cell pictured here hangs between adjacent frames.*

top. More bees in the air, louder sounds, and a generally agitated state typify a short-time queenless colony.

Other factors can also elicit these behaviors. A skunk or raccoon visit the previous evening can agitate the bees for most of the following day. A whiff of pesticide—not enough to kill lots of bees—can cause fanning, agitation, and defensiveness for several days. Opening the colony on a cool, rainy day or during an extreme dearth can cause defensiveness because there are more bees than usual at home. You'll have to explore a bit to be sure of the cause, but the sound of a short-time queenless colony is distinctive. If the colony has been queenless for a week or longer, other behaviors emerge, such as an egg shortage and supersedure cells—a sure sign of what's going on.

If the behavior continues for longer than a month, workers may be laying unfertilized eggs, foraging will be significantly reduced (when compared to other colonies in your yard), and there may be no drones. After a month, a colony can remain queenless if it simply wasn't able to raise a queen. Even if they did produce several supersedure queens, all can be lost while fighting for supremacy (it happens often enough to mention),

or the winning queen can be eaten by a bird while out looking for a mate.

There can be a lot of bumps in the road back to monarchy without your timely intervention.

Emergency Supersedure

One event that requires a colony to produce a new queen is the sudden death, loss, or severe injury of the current queen. Sudden death can occur if the queen is accidentally crushed by a beekeeper during a routine colony exam. Loss can also occur when a frame is removed, and startled by the sudden exposure to light, the queen flies off, looking for the warm and dark broodnest from which she was suddenly removed. Not a strong flyer, she can become lost, sometimes not far from home.

Any injury is likely to alter the queen's ability to lay eggs, produce pheromones, and eat. These deficiencies are immediately evident to the workers because of the constant attention the queen receives. They may or may not continue tending her, and in a time as short as one to three days, they react to the reduction or lack of egg-laying, queen substance, and her other pheromones.

These events signal the beginning of queen-replacement behavior among the workers. Because of the urgency, this process is referred to as an emergency supersedure.

During swarm preparation, the colony receives a series of signals and reacts to each in turn, building up to the finale. Unlike during swarm preparation, workers don't make queen cell cups during an emergency supersedure, because no queen is available to lay eggs in them. Instead, the house bees and those actively feeding the brood search for eggs or the youngest larvae they can find that are still feeding on royal jelly—the special diet fed to future queens, which allows complete development of their reproductive organs.

When eggs or royal jelly fed larvae are located, the wax builders begin building queen-size cells for them. Because the egg or larvae could only be found in a regular, horizontal cell on a frame, the queen-size accommodations are built outside the frame's face and extend down and between adjacent frames. Several of these may be made by the colony if resources, such as food and larvae of the right age, are available.

Normal Supersedure

The second event that can trigger queen replacement is the normal aging of a queen. As the queen ages, the sperm she acquired when mating is slowly depleted and eventually is gone. As this occurs, she lays more and more drones (unfertilized eggs turn into drones) and fewer workers, creating an unbalanced hive population. Though a healthy, prosperous colony can afford to support a large population of drones, it is the workers that ensure colony advancement and survival.

To regain the balance, a fertile queen that produces mostly workers is needed. The colony will produce another queen in the same way as in an emergency supersedure, but with the insurance

A frame that has drone comb scattered all over can be the work of a drone laying queen, or it can be in a colony that has laying workers. In either case, there is insufficient worker brood, and the colony is generally demoralized and anxious.

You may find several supersedure cells on a frame and several frames with supersedure cells. Or, there may be only one or two supersedure cells in the whole colony. It depends on available resources and the availability of larvae of the right age for workers to raise as queens.

of the presence of the current monarch. The first queen to emerge usually destroys those not yet emerged, leaving her in charge. Often, however, she and her mother will remain in the colony, both going about their queenly duties—laying eggs. If the old queen is still producing some workers, the colony enjoys a burst in population. Eventually, the older queen expires, leaving her daughter the sole monarch.

Drone-Laying Queen

Occasionally, you will have a queen that you have just purchased—or even one you have had for some time—that lays mostly unfertilized eggs, which produce drones. This can happen if a queen was not mated, or was poorly mated because the queen producer did not have enough drones to mate with the queens, or if the weather during the queen's short window of opportunity for mating did not allow her to fly to drone congregation areas.

She will appear normal, and if she is accepted by the colony, she will begin laying eggs. She lays them in regular worker cells, but none of them are fertilized; therefore, they all produce drones. This is very confusing for the colony, and also for you, initially.

It will take seven to ten days to recognize the situation, which is a great waste of time for the colony because this queen needs to be replaced immediately. This imbalance can also happen when an older queen eventually depletes her store of sperm and is unable to produce fertilized worker eggs. This is usually first noticed in the brood area. The usual solid pattern of closed cells will have open cells in places and a few drone cells scattered throughout the frame, instead of along the edges, as is the normal location of drone cells. It occurs gradually, over two or three weeks, so you should notice the increase and order a replacement queen. The colony, too, usually recognizes this condition as a failing queen because the population becomes unbalanced, signaling a series of behaviors resulting in supersedure. To prevent a lapse in laying or a battle among emerging queens, look for supersedure cells and remove them before introducing a new queen that you have ordered.

Laying-Worker Fix

A variety of mishaps—old age, injury, diseases, mites, exposure to pesticides—can befall a queen and cause her to stop laying eggs. Normally, a colony will note this change and begin preparations for producing a new queen: an emergency supersedure. Critical to this event is the presence of eggs, or larvae that are three days old or younger. These very young larvae have had only royal jelly as a diet. Older larvae have been switched to the less nutritious worker jelly, causing permanent physiological changes in their development. But sometimes communications break down, and the message that the existing queen is malfunctioning doesn't make it to the workers. By the time the problem is discovered, if at all, there may be

Unite a laying-worker nucleus colony, on top, with a strong, healthy nucleus colony on the bottom. Remove the excess paper from the outside and wait a few days. The bees will (almost always) sort this out themselves, leaving a single colony with one queen and no laying workers.

no eligible larvae or eggs available, and the colony cannot by itself produce a new queen.

Without the queen's regulating presence, the ovaries of some workers begin to develop, and they gain the ability to lay eggs. Because they have no capacity to mate, all of the eggs they lay are unfertilized and will develop into small but functioning drones.

The eggs are laid in regular worker cells. Because a laying worker is smaller than a queen, many of the eggs she lays do not reach the bottom of the cell and cling to the sides. And because there are several, perhaps many laying workers, you may see several eggs in a cell. Other workers remove multiple eggs and raise a single drone larva in each cell. Initially, the overall pattern on a brood frame will be spotty, with some cells unoccupied, some with multiple eggs, some with normal-looking larva, and perhaps some capped drone cells. It is a confusing mess. Left to its own devices this colony is doomed, and if it's your colony, intervention is necessary.

It takes a colony several weeks to reach this sad state, and experienced beekeepers, rather than invest the time and effort required to save the colony, simply let it expire. This choice becomes obvious late in the season when even heroic efforts usually prove futile.

However, if your laying-worker colony is discovered early in the process, or early enough in the season, there is a good chance it can be saved. Here's how to combine it with one of your other colonies. Reduce the laying-worker colony to one or at most two broodnest boxes. By now the colony is weak, so combine frames from the two or three brood boxes into one or two boxes. Put most of the brood and as many bees as possible into a single box.

To do this, first remove any empty frames from the box into which you are going to put these frames. Then, remove the frames that have brood in them, along with the adhering bees, and fill the empty spaces. Take the rest of the frames and shake the bees into the new boxes.

Then, remove the cover, inner cover, and any honey supers on a nearby strong, healthy colony (with a queen). Place a sheet of newspaper over the top of the frames, and using your hive tool, make a slit in the paper between two or three frames, removing the excess from the edges. Place your laying-worker boxes directly on top of the newspaper, replace any honey supers above this, and close up the colony. This is called the "newspaper method" of uniting colonies. It is easy and generally successful.

The bees from both sides gradually remove the paper (in a day or several days) and carry it outside. In the process, the chemical messages from the queen below, coupled with the multitude of bees from the colony below that begin streaming up, essentially overwhelm the addition. The queen's pheromones spread throughout both colonies and begin to bring the two together. At the same time the laying workers are affected by the queen's pheromones and slow down or stop laying. They generally don't last very long anyway. After a week or so, the union is as complete as it is going to be, and where there were two, now there is one colony.

Preventing this extreme measure and the resultant colony loss is, certainly, less work and less expensive. Colonies that are queenless almost always have some distinct, noticeable behaviors that provide clues to the situation.

Broodnest

A broodnest in a typical colony is oval shaped. This is where the queen deposits her eggs, the larvae are fed, and the sealed brood is kept. Though the volume of this football-shaped region changes during the season as bees enlarge it, the shape is relatively constant.

The cells in the top two-thirds or so of the broodnest have pollen placed in them by returning pollen-collecting foragers so the pollen is close to the young bees that need this food.

Nurse bees, which need pollen in their diet to complete development, spend their first few days in the warm, safe center of the nest, close to the food. They also need pollen to produce the glandular food they feed to the developing brood. Keeping pollen close to the broodnest is a matter of efficiency and necessity.

Surrounding the pollen ring, on the sides and above the broodnest, the bees store ripened honey. The honey stored closest to the brood is continuously used for food and replenished. Above the broodnest, the bees store the surplus honey needed to feed the colony during the part of the year when plants are not producing nectar. The temperature in the central part of the broodnest, when brood is present, is held at a constant 95° F (35° C).

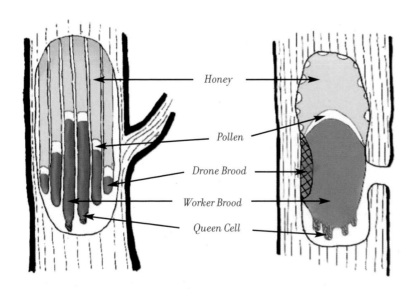

Honey

Pollen

Drone Brood

Worker Brood

Queen Cell

Two views of a typical hollow tree cavity showing the arrangement of honey, pollen, and broodnest. Queen cells are produced on the very bottom of the broodnest, where the wax is the newest.

🐝 *The queen lays eggs in the broodnest. You can see them in the bottoms of these cells.*

🐝 *A typical broodnest frame has honey at the top and brood in the center. Between the brood and the honey is a narrow band of pollen.*

The design of the broodnest serves several purposes. When the colony is young and small, the broodnest begins close to the top of the nest. As the nest expands, the broodnest area migrates toward the bottom, following the expansion.

At the end of the growing season, when brood rearing slows or ceases and nectar and pollen are no longer available, the living arrangement changes. Without a large brood area to protect and keep warm, the bees stay close to the larger mass of stored honey above and to the sides of the broodnest. They continue to move in the nest until they run out of food (and subsequently starve) or until nectar and pollen again are available.

During this time, the queen may continue to lay eggs, but she tends to follow the cluster of bees as they move up the nest. If she ceases laying, she stays with the cluster. If the nonproductive season is not severe, the broodnest remains in its original location, because food is easily obtained.

Understanding this pattern of movement and how the bees construct their nest is important in managing a colony over the seasons. Anticipating what the bees are going to do allows you to prepare adequate space for them to move into. Replacing the combs of the broodnest after three or four years of use is encouraged to provide brood and young bees with new, clean wax.

The Workers

During the active season, a typical honey bee colony contains a single female queen, a few hundred drones, and thousands of female workers.

Workers raise the young, build the house, take care of the queen, guard the inhabitants, remove the dead, provide metabolic heat when it's cold and air-conditioning when it's hot, gather the food, and accumulate the reserves needed to survive the inactive season. When all goes well, workers also provide a surplus of honey for the beekeeper to harvest.

Queens and drones are fairly task-specific for their entire lives. What makes workers so interesting, and so complicated, is that their tasks change as they age, yet they remain relatively flexible and can switch between tasks when needed.

A worker starts as a fertilized egg, with half of her genetic traits taken from her mother, the queen, and the other half from one of the many drones with which her mother mated. She

The worker bee that appears in the top center of this picture has her head in a cell, feeding the larva inside. Note the two bees in the bottom center of the photo. They are transferring nectar from a forager to a food-storing bee. This is part of the nectar-ripening process, which occurs in the broodnest on what is often referred to as the dance floor.

🐝 *Worker bees are smaller than both queens and drones. They are by far the most numerous bees in a colony. The queen and workers are shown here for comparison.*

◀▌)●◀

Worker eggs in brood cells. To find worker eggs, start by looking at frames from the center of the broodnest. Carefully remove the frame, because the queen may be on it and you don't want to injure her. Hold the frame in front of you, with the Sun shining over your shoulder, straight down into the bottoms of the cells. The eggs are small, white, and centered on the floor of the cell. They are just a bit smaller than a grain of white rice. Finding eggs takes practice, but it soon becomes second nature.

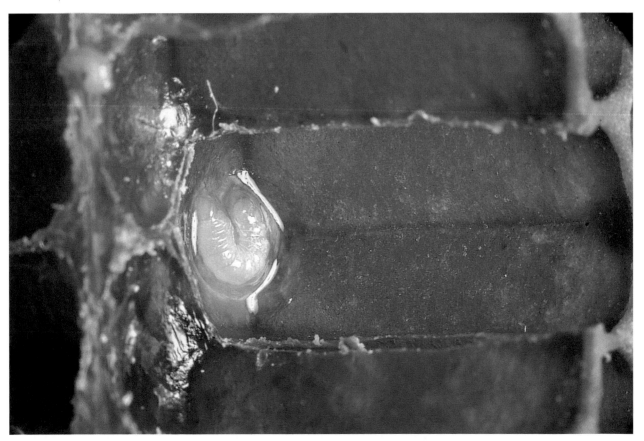

🐝 *After three days, the eggshell, or chorion, dissolves and a tiny larva emerges. The larva lies flat on the floor of the cell and is fed thousands of times a day by house bees. Like a queen larva, it floats in royal jelly. After three days, however, its diet is changed to worker jelly.*

ABOUT BEES

Honey Ripening

Nectar collected from flowers is roughly 80 percent water and 20 percent sugar. Though other sugars are present, sucrose, a twelve-carbon sugar molecule, is the predominant sugar. The sugar content of nectar varies depending on the environment, the age of the flower, and other factors. During the flight home, the forager adds an enzyme called invertase to the nectar to begin the ripening process. Adding the enzyme changes the twelve-carbon sugar to two six-carbon sugars: glucose and fructose.

When a forager returns, she gives the nectar to a receiving house bee. This bee first adds additional invertase, and then finds a location in the hive where she can further tend to the nectar. If the rush of incoming nectar is hectic, such as during a heavy nectar flow in the busiest part of the day, she will place the nectar in an empty cell or perhaps in a cell with a small larva. The droplet will hang from the ceiling of the cell, exposed to the warm air of the colony until moved later.

Eventually, the nectar, which has been acted on by the enzyme and evaporation, is reduced to a mixture that is 18 to 19 percent water and just over 80 percent sugar, or what we call honey. Individual droplets are collected when ripe and placed in cells. When a cell is full, it is covered with a protective layer of new beeswax.

When nectar has been ripened into honey, it is stored in cells in the broodnest area or above the broodnest area in the surplus honey supers. When a cell is filled with honey, house bees cover the cell with new beeswax for protection.

A foraging honey bee trying to gain entrance to a colony other than her own will be inspected by attending guards and, if found to be in the wrong place, will be physically removed. Several guards may join the fray if required. This system is imperfect, however—if traffic is heavy, a foreign forager laden with nectar or pollen may be allowed inside. The same thing happens when drones are returning from mating flights in late afternoon.

emerges from the egg as a larva, and for the first three days, she is fed a diet identical to that of a queen larva. After that, her rations are cut (see progressive provisioning on page 33). As a result, her reproductive and some glandular organs do not fully develop. She is not as large as a queen, is incapable of producing queen substance, and is unable to mate. After three days as an egg, six days as a larva, and twelve days pupating, she finally emerges as a fully formed female adult worker honey bee.

Like any newborn, the first activity of this new worker is eating. Initially, she begs food from other bees, but soon she begins to seek and find stored pollen. This high-protein diet allows her glands to mature for future duties. She stays close to the center of the broodnest—the warmest and safest part of the colony— which is also where most of the pollen is stored. Within a day or so, she joins the labor force, learning increasingly complicated tasks as she matures. She begins in the broodnest, removing debris from vacated pupa cells, pulling out the cocoon and frass (waste) that can be removed. Others follow her, polishing the sides and bottom of the cells with propolis in preparation for another egg.

After a few days, the glands in her head (the hypopharangeal and mandibular glands) are nearly mature, and the worker begins feeding older worker larvae a mixture of pollen and honey. As these glands mature, she can feed hatchlings royal jelly produced in those glands. She is also able to feed the existing queen this glandular food. At the same time, she can groom the queen, help remove waste, and pick up after her. As she works, she picks up and distributes tiny amounts of queen pheromone throughout the colony, assuring all inhabitants that all is right in the world— or informing them that all is not right, that the queen may be failing, or even missing. This activity is all important in maintaining the status quo in a colony or in initiating a change that will again bring balance to the colony.

After a few days of cleaning, feeding, and eating, this worker begins to explore the rest of the nest, traveling farther and farther from the center. Soon she ventures near the hive's entrance and begins taking nectar loads from returning foragers—the first step in turning nectar into honey.

If there is sufficient room in the hive for nectar storage, the workers continue to take it from returning foragers. As room becomes scarce, however, the workers become reluctant to take it. Foragers with average- or less-than-average-quality nectar can be turned down as the number of bees visiting high-reward flowers increases. The same goes for pollen collection, which is influenced by the number of young needing to be fed.

Thus, if there is plenty of storage room and enough house bees available to take the returned booty, foraging increases. In fact, more and more foraging-aged bees begin to forage, and more and more receivers are recruited, leaving fewer house bees to clean and feed the young, as well as fewer guard bees to watch the hive. All of these factors spur the colony to collect nectar and pollen at an astonishing rate.

These older house bees also take on other tasks as needed, such as house-cleaning duties—removing dead bees, dead larvae, and debris, such as grass and leaves. (For more information, see Foragers, on page 51.)

Guards

After two or three weeks, a worker's flight muscles are developed and she begins orientation flights around the colony. Even before this, however, the glands and muscles of her sting mechanisms have matured, and she is fully capable of defending the nest. Therefore, she becomes a guard. In a large colony in midseason, the number of dedicated guards at any one time is

relatively small—maybe 100 or so. However, if there is a large threat, thousands of bees can be recruited almost instantly. These new guards are temporarily unemployed foragers, older house bees, and resting guards.

Guards perform multiple tasks. They station themselves at the colony entrances and inspect any incoming bee. This inspection is odor based, because bees have a distinct and recognizable colony odor. If a forager returns to a colony that's not hers, she will be challenged at the door.

Other insects are also challenged if they try to enter. Yellow jackets, for instance, may try to help themselves to a colony's honey. When this happens, the thief is met by several guard bees that wrestle and struggle with the intruder. They will bite and sting the intruder attempting to kill or drive it off.

Animals that try to steal from a colony are also rebuked. Skunks, bears, raccoons, mice, opossums, and even beekeepers will be challenged, threatened, and eventually attacked. When confronted by a large intruder, such as a beekeeper, some guards will engage in intimidating behavior before stinging. They will fly at the intruder's face (they are attracted to the face because of the eyes and expelled breath) without stinging. This action can be annoying but—if the beekeeper wears a beekeeping veil—inconsequential. If such warnings fail to drive off the intruder, more guards will be attracted to the intruder. If the intruder's attack on the hive continues, the bees will sting.

When a honey bee stings, her sting pierces the skin of the intruder. The sting is a three-part apparatus, made of two barbed, moveable lancets and a grooved shaft. The lancets are manipulated by muscles. The shaft is connected to the organs that produce the venom and acid that are injected into the skin.

After the sting is imbedded, these muscles contract, relax, and contract continuously. Each contraction pushes a barbed lancet further into the skin with the venom gushing down the shaft.

Because the lancets are barbed, the bee cannot extract them. When she makes her escape, the sting apparatus is torn away, remaining in the victim's skin. This is seldom a slow, methodical process. Guards approach an intruder, land, sting, and escape in

Cleaning Protective Gear

After you have worn your bee suit and your gloves for several colony examinations, the amount of venom and alarm pheromone begins to build up in the material. Frequent washing will eliminate these chemicals, and reduce visits from guards when you work a colony. Wash these clothes in a separate load so that alarm pheromones don't contaminate your other clothes.

Throw Off the Scent

When working a colony, you may inadvertently kill a bee, which can also release some alarm pheromone. Other bees will notice and respond. Smoking the colony masks the alarm pheromone, reducing the number of guards that respond to the alarm. If you are stung while working your colony, you are marked, but you can reduce the response of other guards by quickly removing the sting apparatus from your skin. Scrape or pull it out, then puff smoke on the sting site. This remedy helps reduce the attack but is a less-than-perfect solution.

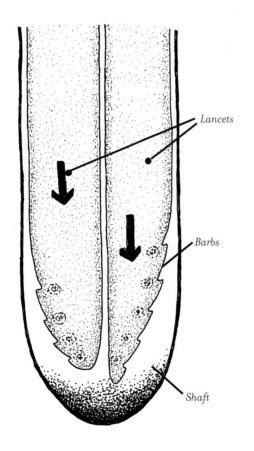

Lancets

Barbs

Shaft

The two lancets of a bee's stinger are barbed and work independently, but in unison, as they push deeper and deeper into the skin of the intruder. The shaft behind the lancets funnels venom into the wound that the lancets are producing.

seconds. You seldom see the bee that leaves her mark. When the guard bee stings, she is mortally wounded. She may, however, continue to harass the intruder. You may see one or more of these bees, when working a colony, with entrails hanging from the end of their abdomens.

When a bee stings, she also leaves behind an alarm pheromone, which alerts the whole colony that an intruder dangerous enough to sting is threatening it and serves as a call to arms. It marks the intruder, enabling other guards to home in on the sting site and further the attack. If the intrusion continues, the number of guards recruited increases until many, many reinforcements are in the air. This increase in guards usually drives off the intruder.

Guard bees make sure they are successful in thwarting your intrusion by chasing you as you leave the colony. This behavior is variable, however. If there is a nectar flow occurring, with many bees coming and going, and the weather is cooperative, guards will seldom follow you farther than 12' (3.7 m). However, the same guards may follow you much farther if there is a dearth of nectar or if the weather is cool and cloudy.

You can often confuse these followers by walking into a stand of tall shrubs or brush, or stepping out of the line of sight of the colony for a moment—behind a building or into a shed or garage. The guards should quickly lose interest.

If you are still being harassed, keep your veil on until they head back home. Smoking these bees does little or no good in deterring their behavior, because they are following you visually as well as by odor. If this behavior is common in your hive, requeening with less defensive stock is recommended.

Robbing

Honey bee foragers have a fundamental goal—finding food. Most often that food is nectar or pollen from flowers. Other food sources can include floral food, sugar syrup in a feeder, or honey found in another hive. Small, weak colonies with few guards are unable to defend themselves from an onslaught of foragers from other hives, but they will try. Guards will fight to the death to keep strangers from stealing their hard-won stores, but if they are overwhelmed, the colony will be robbed of all of its honey, and in the process, many of the bees will be killed.

Beekeepers can inadvertently expose a weak colony to robbing by opening it during a dearth, or at any time foragers from other colonies are having difficulty finding food. Opening the colony sends an aromatic message—food!

There are several signs that a colony is being robbed, and it pays to recognize them before the colony is destroyed. There will be many bees at the entrance fighting, with workers balling up together with five, six, or maybe ten in a ball. There will be individuals rolling on the landing board and falling off, and all the time more and more bees arriving at the colony. Bees from one, two, maybe all the rest of the colonies in your yard can become involved.

Because of the mayhem and fighting, alarm pheromone fills the air. You may even smell the banana-like odor. Guards rush out of their colonies searching for the source of the alarm pheromone but will be unable to locate a typical intruder. When this happens, they can become defensive in a hurry, stinging everything and anything for several yards in all directions. A rob-

Honey robbing by outsider bees can begin during a beekeeper's inspection, when honey is exposed. If robbing begins, close up all colonies as fast as you can. Restrict the entrances of the colonies doing the robbing by inserting grass or a reducer, and completely close up the colony that is being robbed until the behavior ceases. A weak colony can be killed by robbing bees during the melee.

bing situation can become fatal for the robbed colony and dangerous for people and pets in the area. If you suspect the colony you are working, or perhaps one you just finished, is being robbed, you have a responsibility to protect that colony before it succumbs. Immediately reduce entrances to all colonies in your yard using entrance reducers or even handfuls of grass. Apply smoke to every colony to disorient the inhabitants and disrupt their rush to rob. Close up the colony that is being robbed, making sure that upper entrances are closed. Seal off the front door with a reducer and grass. This stops outside bees from entering and allows the colony being robbed to reassemble its forces.

Bees that were robbing will continue to investigate previous openings, trying to gain entrance. If you've done your job, they won't be able to enter and in anywhere from a few minutes to a few hours, the turbulence in the area of the colony subsides. When it appears that life is back to normal, it will be safe to slightly open one entrance to the colony that was being robbed. Leave the entrance reduced for days, or even weeks if the colony remains small or weak. Don't work *any* colony during a dearth, and know and watch for signs of robbing. It may be the worst thing that can happen to your colony and to your good reputation as a beekeeper.

 Dandelions are nearly universal and provide much-needed nectar and pollen early in the spring when they bloom.

Sweet clover is one of the dominant nectar plants in the northern hemisphere.

Foragers

When a worker matures and ventures outside the colony on a routine basis, she becomes a forager. This period of her life starts when she is about three to four weeks old, but may be sooner if the colony needs foragers. She may be a scout bee, seeking new sources of nectar, pollen, water, or propolis; then collecting some and returning to the hive to share her newly found information. Or, she may be recruited by another scout bee or forager to visit a particularly productive patch of flowers.

If food is dispersed evenly, foragers exploit nearby areas, circling the colony. This is seldom the case, however, because flowering trees, shrubs, and weeds grow where they can, rather than where bees would prefer them. And as the season progresses, bees' tastes change. Therefore, the forage area changes from day to day. To add to this complexity, some plants produce flowers, nectar, and pollen during only part of the day. Cucumbers, for instance, bloom from very early in the morning until about noon, when their blossoms wither. Bees visiting a cucumber patch learn the daily schedule and visit during the morning only. They may turn their attention elsewhere in the afternoon or take the rest of the day off.

As mentioned earlier, workers forage for nectar, pollen, water, or propolis, but not all at the same time. Some foragers gather nectar only and continue in that work for as long as the nectar is received back home. Some collect only pollen. Others, however, collect both during the same trip. We'll examine foraging for water and propolis later, but the relationship between bees and flowers is not only fascinating but critical to the success of the colony.

Finding food is the job of scout bees. Experienced scouts seek food using the color, shape, markings, and aroma of flowers. They learned that particular flower shapes, colors, or aromas signaled a reward and return often, or look for similar signs elsewhere. Beginners may recognize a familiar aroma, learned when they were in the hive working as a food storer, and investigate.

When a scout locates a promising source, she investigates its value. She lands on a flower, if it is large enough, or on a nearby stem or leaf so that she can reach the flower. She extends her tongue, called a proboscis, folding its three sections together to form a tube, and sucks in the nectar. She may scrabble in the center of the flower or brush against anthers in her pursuit of nectar, gathering pollen on her body hairs.

When full, she leaves the flower, circles the patch a few times—to get her bearings by noting landmarks and the position of the Sun—then heads for home.

Because the forager has found what—to her, at least—is a new patch, she usually tries to recruit other foragers to visit the patch. She initiates the dancing behavior on the comb in the lower part of the broodnest where other foragers gather, either waiting to be recruited or offloading pollen and nectar from a recent trip.

At the same time, the house bees who meet the scout on the dance floor have a message of their own. If there is a shortage of nectar in the colony and room for new stores, the house bees will unload the scout almost immediately. If, however, there is no room, house bees can refuse incoming nectar, effectively shutting down foraging activity. A critical situation confronts the colony at this point: If room is limited, yet there is a strong nectar flow on, they may decide to place nectar in the broodnest, where the queen is busy laying eggs. Doing so reduces the space in which the queen can lay eggs, and may initiate swarming behavior, and in extreme cases, this may lead to nest abandonment if there is not enough room for food storage and brood rearing to assure survival.

Other food-storing bees work with the pollen brought in by foragers, who dump it in a cell near the broodnest. Young workers pack it into the cell, using their heads as rams, until the cell is nearly filled. They leave a shallow space at the top of the cell to be filled with honey, which acts as a preservative for long-term storage.

As hazardous as being a guard bee may seem, it doesn't hold a candle to the dangers encountered by a forager. After graduating from the home duties of feeding, cleaning, and guarding, a mature worker bee is able to wander far and wide in search of food. When out in the field, a lone honey bee can fall prey to birds, spiders, preying mantises, and an array of other predators. The weather, too, works against her, with sudden showers, rapid temperature changes, or high winds making flying difficult. Other dangers include rapid automobile traffic and even flyswatters.

One danger that can threaten nest mates as well as the forager is insecticides. When the forager comes in contact with flowers that have had an insecticide applied to them, she will probably die almost instantly. Worse, she may harvest contaminated nectar or pollen and return home with it. If it is nectar, she will share this lethal cocktail, causing the death of others as it is moved throughout the colony. If she carries contaminated pollen, she may die, but not before she stores it. Later, this lethal poison will be fed to developing larvae, the queen, or nurse bees.

If the forager avoids these dangers, old age will finally claim this five- or six-week-old bee. Foraging is the most personally expensive (excluding, of course the supreme act of defense— stinging) behavior of honey bees. Muscles deteriorate, body hairs are pulled out, and wing edges become frayed as the forager's body ages. Too tired or slow to make the flight back to the hive or to escape a predator's attack, her short, purposeful life ends.

Communication

Something we don't often think about is that, with the exception of a very small area near the front door, the entirety of the hive's interior is pitch black. There are no windows or skylights, and any small cracks are sealed with propolis. Everything that goes on inside is done by touch, feel, and smell. Bees don't see each other inside, nor do they see the cells to clean, the larvae to feed, the dead to remove, or the honey to eat. Yet, when foragers are out in our world they navigate by light and sight, by color and location.

Therefore, the returning foragers must translate their visual experiences in the outside world to their nest mates using non-visual methods. When a forager discovers a new flower patch and has tested the quantity (to some degree) and the quality of its harvest, she returns to the nest to advertise its location and potential.

When she returns to the hive she goes to an area in the broodnest close to the main entrance. There, foragers not yet foraging or those that finished earlier congregate. Also there are those bees who take nectar from incoming bees, other foragers who have returned and are being unloaded, as well as those very young bees that are cleaning cells and caring for the young.

To advertise the patch, the returning bee begins the famous waggle dance, which indicates with some—though not exact— precision the location of the source of her harvest. The information includes the distance (measured in expended energy during the trip home) and direction (where the Sun is in relation to the colony and the patch). The value of her find is communicated to others by the intensity and duration of her dance. Unemployed foragers will come close to taste and smell the nectar, and some (unemployed foragers) will follow her dance through several performances and eventually leave the hive in search of the source. Remember, though, there are many, many foragers recruiting simultaneously on this dance floor, but unemployed foragers do not sample each dance. Rather, they pick one and follow. They don't follow several, evaluating and comparing the differences.

The accuracy of the dance in pointing additional foragers to the patch is fairly reliable but not infallible. Factors such as obstacles (tall buildings) and head or tail winds enter into the equation as well. But outgoing foragers also have the scent of the floral source to help guide them, and a downwind approach can assist in locating the patch. Even so, many recruits leave the dance floor to find this floral patch, only to return in a short

Metamorphosis

Honey bees undergo what entomologists call complete metamorphosis. Complete metamorphosis describes the maturation of an insect from egg to adult. Because insects have hard exoskeletons, they cannot increase their size or the number of internal organs at will, so they produce a skin, grow into it, shed that skin, and produce a larger one. They will do this several times until they are as large as they can grow. Each of these stages is called an *instar*. Honey bees have five instars. In the last instar they cease feeding and produce a thin, silklike cocoon that covers the body. House bees, cued to the change when a larva quits eating, cover the cell with a mixture of both new and used wax. In 12 days, the transformation is complete, and a new adult pushes and chews her way free of her youthful confines. Her metamorphosis is complete.

(1) An egg is pictured standing on end, held there by glue used just for this purpose. In the cell to the right of the one containing the egg is a first instar larva, already floating in royal jelly fed to it by house bees.

(2) Larvae grow rapidly, going through five instars in six days. Here are two different instars (development stages between molts).

(3) When the larvae are ready to pupate, they stand upright in the cell, stop eating, void their digestive systems into the bottom of the cell, and prepare to spin their cocoons. Noting this change, house bees begin covering the cells with a mixture of beeswax and propolis. Now is the time that female varroa mites enter the cells to parasitize the larvae. (Varroa mites are discussed in the next chapter.)

(4) Several stages of pupating workers are shown here. The wax cappings have been removed to show the developmental stages. At top left is a pupating worker nearing maturity, whose eyes have already developed color. At the center of the bottom row is a worker nearly mature enough to emerge as an adult.

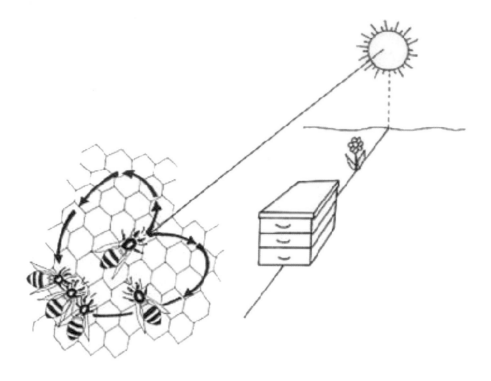

A simplified diagram of the waggle dance, which conveys information on the location of a food source in relation to the hive and the Sun. The actual dances are much more complicated.

time to try again. They didn't get all the instructions, it seems. Those that find the patch, however, will return, and they too will recruit additional foragers if they found it to be profitable. You can see that this communication allows a colony to exploit many patches simultaneously, and that better, more rewarding patches will be highly recruited whereas smaller or less rewarding patches will be abandoned.

At the same time, the colony must adjust its capability to accommodate this influx of nectar. Returning foragers will actually recruit nonforagers to become food-storers by performing what is called the tremble dance.

Coordinating intake and storage as efficiently as possible allows a colony to quickly exploit as much of a good nectar source as possible. Moreover, it allows the colony to adapt to a changing environment to best exploit new sources, and it minimizes the time individual foragers spend searching.

Pollen—Pure Flower Power

Pollen is produced in a flower's anthers as part of the reproductive process. When mature, the anthers dehisce, shedding their pollen. Individual pollen grains are transferred to the stigma of a receptive flower, which, depending on the species, can be the same flower, different flowers on the same plant, or flowers on different plants. Pollen travels into the female part of the flower—the ovary—and produces the seed and the endosperm surrounding it. (Think of apple seeds and all the rest of the apple surrounding the seeds as the endosperm.) Pollen transfer is accomplished by wind, moths, butterflies, bats, birds, and a great variety of nectar- or pollen-feeding insects, including honey bees.

In their quest for nectar, honey bees come into contact with pollen on a routine basis, because flowers don't produce nectar until the pollen is mature. Some plants, such as cucumbers, produce male flowers that have both pollen and nectar but no ovaries for seed production, and female flowers that have nectar only and can produce seeds. To accomplish pollination, a honey bee visits both flowers (floral fidelity), receiving nectar from both and pollen from the male flowers. She then shares that pollen with the female flower during her visit.

Pollen grains have a minute negative charge, and bees have a minute positive charge and thousands of multibranched hairs capable of attracting, capturing, and holding pollen grains. Foragers clean most of these pollen grains out of their hair using their legs and carry them home packed in the corbiculae, or pollen baskets, on the outside of the hind legs. But while the bee visits other flowers, some pollen is transferred, and the plant has accomplished its goal.

Pollen is the only source of protein, starch, fat, vitamins, and minerals in a colony's diet. By weight, pollen has more protein than beef, and is *the* best food for developing larvae and young adults, and for producing brood food.

A colony will collect nectar and pollen from thousands of plants daily, and from hundreds of different plant species during the course of a season. This diversity provides a balance of the essential nutrients from pollen needed to grow healthy larvae and maturing house bees, and a rainbow of stored pollen inside the hive.

Collected pollen is returned to the hive by the forager, who promptly dumps it into a cell that already contains some pollen. House bees pack the pollen into the cell tightly so space is carefully used. Pollen is stored near the broodnest, where it is used at an amazing rate. Sometimes, an extraordinary amount of pollen will be collected, and entire frames will be filled with this multicolored food. Stored pollen fills only a part of a cell and is covered with honey for preservation and later use.

When bees visit blossoms, they pick up pollen and carry it back to the hive.

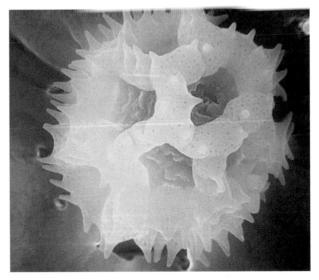

Every plant's pollen is distinct in shape, markings, color, and nutritional value. This is sweet clover pollen, greatly magnified.

🐝 *Pollen loads are taken into the hive by the forager, who then puts her hind section into an empty cell near the broodnest or one partially full of pollen, and kicks off the pollen loads (often called* pellets *in beekeeping literature). The cells are left partially full; later honey will be added on top as a preservative.*

🐝 *When pollen is abundant, a colony will often store nearly entire frames of it. An overwintering colony can use two to three full frames of pollen to feed brood in the spring when brood-rearing starts, before pollen-bearing plants are blooming.*

Propolis

Honey bees use propolis to stick things together to keep them water- and windproof. A cover, such as this one, can be difficult to remove when it's warm outside because the propolis will be sticky. When it's cold outside, the propolis will be brittle—when you open the cover, it will vibrate and snap, an activity the bees inside frown upon.

When routinely working your colony, take a moment to clean the frame rests and frame ends of sticky propolis. If left too long, the buildup will cause the frame tops to violate bee space. Also, clean frames help keep your hands and clothes clean when working the colony.

Honey bee foragers collect nectar, pollen, water, and propolis—a wonderfully mysterious substance. Most plants have evolved some form of self-protection from microbial, insect, or even animal predation. Thorns, stinging hairs, bitter flavor, and poisons all are used by plants to thwart being eaten. A technique used by some species is the exuding of a sticky, microbially active resinous substance that covers leaf or flower buds while they are developing to protect them during this tender stage. Other plants secrete resinous substances around wounds for the same protection.

Bee foragers collect this substance by scraping it off buds or wounds with their mandibles and packing it away in their pollen baskets. Because freshly collected propolis is sticky, other bees help remove the mass when the forager returns to the hive.

When propolis is warm, it has the consistency of chewing gum and will stick to anything it touches.

Propolis varies in color, depending on the plant from which it was collected, ranging from nearly black to brown, red, or gold. Also, the various microbial chemicals vary, as do the other compounds. In spite of this variability, however, no matter where it is collected, propolis is more similar than different, suggesting that bees seek plant resins with a detectable quality. The protective aspects of propolis should not be underestimated: it is active against potentially pathogenic bacteria and fungi in a hive.

Beekeepers may inadvertently combine very small amounts of propolis with beeswax, lending both color and fragrance to a burning candle. Because of its antimicrobial properties, propolis is also collected by some beekeepers and used in simple ointments, salves, and lozenges.

Honey bees will cover objects too large to move in the hive—such as these two mice—with propolis. The propolis retards bacterial and fungal growth and keeps the bodies from rotting. They just dry out when covered with this material.

Water

Foragers collect water where they find it, when they need it. Ponds like this one are ideal because they have an odor and are easy for other foragers to find.

Water is crucial to honey bee survival. Nectar, which is mostly water, provides much of the needed moisture in a colony but not all of it. Foragers will collect water when there is a need in a colony. Water is used to dissolve crystallized honey, to dilute honey when producing larval food, for evaporative cooling during warm weather, and for a cool drink on a hot day. A full-sized colony at the height of the season will use, on average, a quart or more of water in excess of the nectar collected, daily.

Water isn't stored, like nectar or pollen, but rather is added to honey, or placed in cells or on top bars so it evaporates, cooling the colony in the process.

Foragers seem to seek water sources that are scented: chlorinated swimming pools or stagnant puddles, rather than fresh water, because there is a scent association for other recruited foragers to use as a beacon. Foragers will mark freshwater sources for other foragers using their Nasonov pheromone. To make water accessible to bees, try the following:

> Float pieces of cork or small pieces of wood in pails of fresh water for the bees to rest on while drinking.
> Install a small pool or water garden, or have birdbaths that fill automatically when the water runs low.
> Set outside faucets to drip slowly (great for urban beekeepers), or hook up automatic pet or livestock waterers.

Drones

Drones are the males in a honey bee colony. As such, they are different from workers and the queen in their physical makeup, their activities, and their contributions to the colony.

A normal colony will produce and support a small number of drones during the growing season. In a full-sized colony at mid-season, as many as 1,000 drones, in all the stages of development, may be present. Drones produce their own pheromones that are recognized as part of the general aroma of the colony.

Drone honey bees emerge from an unfertilized egg laid by the queen, which means that the genes carried by the drones come from *only* the queen. The cells in which drones are raised are a bit larger than worker cells and, like worker cells, are part of the comb on a frame rather than hanging below it or butting between combs, where queen cells are constructed.

Drone cells are almost always located along the edge of the broodnest area, often in the corners of a frame. This placement helps drone development; drone larvae and pupa do best in temperatures just a degree or two cooler than the 95° F (35° C) in the very center of the worker brood area. Drones take twenty-four days to develop from an egg to an emerged adult. They spend six-and-a-half days as a larva, fed a diet that is a bit more nutritious than a worker's diet but not nearly as rich as a queen's. Workers, by comparison, are larva for six days and pupa for twelve days. Queens are larva for five-and-a-half days and pupa for only seven-and-a-half-days. A colony invests a lot of food and energy in raising drones.

A drone larva sheds its skin as it grows (as do worker and queen larva). When the process is complete, the workers cap the cell with a mix of old and new wax. Because of the drone's large size, these cappings are not nearly flat, like a worker's, but domed to provide additional room, and are often referred to as bullet-shaped caps.

After 24 days (depending on the broodnest temperature, this time frame may vary by a day or two), an adult drone emerges. They are vastly different in appearance and function from their worker half-sisters. Drones have no sting apparatus. (A sting is part of a worker's underdeveloped reproductive system.) They are larger than workers, have comparatively huge eyes that reach to the tops of their heads, and have a stout, blunt abdomen.

For the first two days or so, they are fed by workers; then, while they learn to feed themselves, they beg enriched food from workers and begin to eat stored honey. After a week of this, they start orientation flights near the colony, learning landmarks and developing flight muscles. When weather permits, they begin mating flights. Drones do not mate with queens from their own hive. They fly to drone congregation areas (DCAs) in open fields, open spots in woody areas, or at the edges of large woody areas. Undisturbed areas serve as DCAs year after year for future

An adult drone can be identified by its large size, and fuzzy, blunt-tipped, stocky abdomen. Note the large eyes, extending all the way to the top of the head. The wings are about as long as the abdomen, unlike the queen's wings, which are only half as long as the abdomen.

generations. Drones tend to be fairly indiscriminate when looking for queens in the drone congregation area (DCA). They will chase nearly anything in the air above of the mating area, such as a stone thrown into the air.

Because drones cannot produce wax, cannot forage, cannot clean house or guard the hive entrance, they are expensive for a colony to support. That investment continues to be borne by a colony to ensure that the genes of the drones' mother—the queen—are carried on in the general population of honey bees.

There comes a time, however, when that price is too high for the hive to pay. If, during the season, a dearth occurs and food income is limited or nonexistent, the colony will, in a sense, downsize its population. They preserve worker larvae the longest and remove the oldest drone larvae from the nest first. They simply pull them out and literally eat them outright, conserving the protein, or carry them outside. If the shortage continues, they remove younger and younger drone larva.

At the end of the season, the colony no longer invests in drones at all. The queen ceases laying drone eggs, (for the exception, see the section titled Drone-Laying Queen), and the workers forcibly expel all or most of the adult drones. Outside the nest they starve or die of exposure.

Seasonal Changes

To begin, imagine three generalized temperature regions: cold, moderate, and warm. The warmest areas are semi- to near tropical; warm regions are moderate, with winter temperatures of around 50° F (10 °C); and cold regions have winters falling to −20°F (−7 °C) or colder.

In late winter, when the days begin to lengthen, the queen begins or increases her egg-laying rate. In warm-winter areas, Italian queens slowed down over winter but probably didn't cease laying eggs. Carniolans and Caucasians probably did stop.

Worker bees consume stored pollen and honey to produce both royal and worker jelly to feed the brood, and as the population increases, so does the need for food. Within a month in warm areas, early nectar and pollen sources become available to supplement stored food.

In the moderate regions, it's still cold in late winter, but not so cold that managed colonies can't be examined by beekeepers on warm days. It'll be a while before good flying weather arrives, but brood rearing is increasing rapidly, and the need for stored food is critical.

By early spring, in the warm and moderate regions, the population is expanding rapidly as the early food sources become plentiful and the weather is increasingly favorable for foraging. In the northern hemisphere, dandelions begin to show in warm areas in February, and populations approach the critical stage. Additional room for expanding brood and food becomes a limiting factor, and swarming behaviors can be observed. Two to three months after the days being to lengthen, swarms emerge in the warmest areas. In the northern hemisphere, this happens in mid-April to mid-May, and mid-May through June in the colder areas, though these times vary depending on the location, management procedures, and weather.

Here, a queen, worker, and drone are pictured. Note that the queen, shown at top, has a long, tapered abdomen. Surrounding the queen are typical workers, which are smaller and have striped abdomens. In the center is a drone. He is larger than the workers. Note the distinctive large eyes.

On a typical brood frame in the broodnest, you can find both worker and drone brood. The worker brood is in the center of the frame, which is the warmest part of the nest. Drone brood is ordinarily along the cooler edges of the frame, especially as shown here, in the top-left corner. Drone brood is capped with large, dome-shaped cappings, which make them easy to spot on a frame. They can sometimes be found on top bars in the broodnest when brood frames are crowded.

Food is abundant in the spring in most warm regions, but as late spring and early summer approach, resources often diminish, and by mid-July they are mostly gone. From then until early fall there is often little forage available, and colonies go on hold, living on stored food (if enough was made during the flowering season, and the beekeeper was not greedy).

In moderate and cold areas, nectar and pollen sources come on strong right after the spring swarming season, and given adequate storage, a colony will collect most of the season's surplus crop starting in mid-June in moderate areas, and early July in the cold areas north of the equator.

In many places in the moderate and cold regions, there may be a slow nectar- and pollen-producing period in midsummer. Commonly called a dearth, this situation can last for a couple of weeks to a couple of months. A prolonged drought is another matter entirely and needs to be addressed as an emergency rather than a seasonal change. If food stores are all used, feeding may be necessary.

By late summer, fall-blooming-plants kick in, and often there is another short but intense collection time. The duration of this collection time, called the *fall flow*, may last for a month or a bit longer, depending on how soon rainy weather and early frosts set in. It's an unpredictable time.

By late November, everywhere north of the equator, most plants have finished flowering, and the colony begins its slow time. Where winter is extreme, the bees cluster in the colony, work to keep warm using metabolic heat generated by vibrating their wing muscles and living on the honey and pollen stored during the past season. There may be extended periods when the bees can't break cluster because of extreme cold. The occasional warm day enables them to move around inside the hive to get close to stored food and to take cleansing flights.

In warm regions, temperature extremes don't exist or don't last very long, and the bees can generally move and fly most of the time, though there may be little or no forage available in winter.

In some areas, such as the American Southwest, the bounty of spring depends to a great extent on winter rains. Early spring plants respond with exceptional flowering and lots of nectar and pollen to harvest. After that, the heat of summer dries up most nectar sources. Late summer rain may bring on a fall flow that holds the colony over until a spring flow begins again. Of course, in all regions agricultural crops can break the rules of nature. Irrigation allows plants to bloom when and where none would normally, providing nectar and pollen sources when they would normally not exist.

As stated, this is a generalized overview—the big picture—of the seasonal cycle of a colony. Plus, we've looked at how a honey bee colony responds to the world, and we've even looked at the individuals in the colony—workers, drones, and the queen—and have a good feel for what to expect from each of them. We need only one more piece of the puzzle—you.

There is so much to wonder at when working with bees that honey and beeswax are, and should be, only one of the rewards. If you can manage a garden in your backyard, you'll be happy to learn that keeping bees follows a similar routine. You need a

commitment in time, but it's not a crippling amount of time. It's doing the right thing the right way at the right time that leads to success. And if you get it mostly right, most of the time, you'll have fun and enjoy the harvest.

Be aware that much of what you will be doing, you will be doing alone, at least when it comes to management. Harvesting and bottling and candle- and soap-making often receive help from interested people in your home, but working bees is probably going to be your single-handed task.

What's Outside
If you've been gardening for a while, you are aware of how your growing season progresses. This is a distinct advantage when you start keeping bees because you are already familiar with the usual benchmarks. Generally speaking, the differences between

Winter Cluster

In the winter, honey bees require protection from the cold temperature and wind. During summer, they work to seal all the cracks in the hive so the interior is relatively windless. Beekeeper-provided wind breaks (evergreens or fences) help as well. Inside, the bees do the rest.

Roughly in the center (top to bottom, sides to sides) of a three-box, eight-frame hive is the broodnest portion of the colony. When the outside temperature drops below 57° F (14° C), the bees congregate in the broodnest area. They crawl into some of the empty cells and fill the spaces between combs. Separated by thin wax walls, this fairly compact mass of bees is called the winter cluster.

The winter cluster is like a ball of bees—the outside layer of the ball is composed of bees tightly packed together and acting as an insulating layer, with their heads facing toward the center of the mass. Nearer the center, the concentration of bees isn't quite as dense, so interior bees can move a bit to get food or care for any brood present.

The temperature required to raise brood is about 95° F (35° C), so the bees in the cluster vibrate their wing muscles to generate heat. As they exercise, their body temperature increases and the generated heat expands outward to help the layer of insulating bees keep warm. These bees, too, vibrate their wing muscles to generate heat and to keep warm. At the very edge of the cluster the temperature reaches but does not fall below 45° F (7° C).

Obviously, the bees on the very outside of this cluster cannot sustain themselves for long at that temperature, so they gradually move toward the center, and the warm, well-fed bees from the inside move toward the outside to replace them.

If the outside temperature falls below 57° F (14° C), the cluster begins to reduce its size, shrinking uniformly. This action reduces the surface area of the ball, and the bees in the insulating layer move even closer together, reducing heat loss from the interior and forming dead air spaces between their bodies. As the temperature drops more, the ball continues to shrink, until it becomes a solid mass of bees. This configuration can be maintained for a short time, but eventually the bees in center, nearest the food, will have consumed all of the honey they can reach.

Bees do not warm the entire inside volume of the colony. All the space around the cluster—above, below, and to the sides—remains at exactly the ambient temperature outside the hive. The only warmth generated is kept in the cluster and is not wasted on empty space. This concentration of warmth is efficient from a heat-conservation perspective, but there is a downside. When the outside temperature becomes very cold, the cluster cannot move to reach more food. If the temperature remains very cold for a long time (less than about 20° F (−7° C), the bees will starve when the food inside the cluster is gone.

When the outside temperature warms, however, the outer insulating layer of bees expands, and the volume of the cluster expands with it, moving to frames with stored honey on the sides of the broodnest or above it.

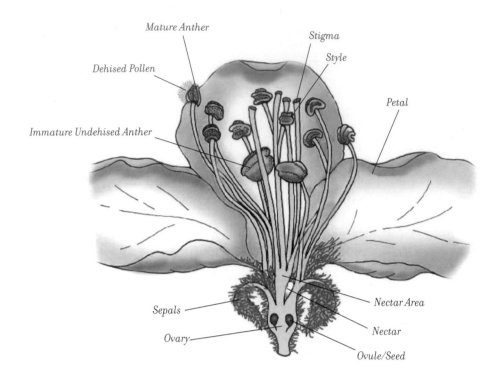

Mature Anther

Dehised Pollen

Immature Undehised Anther

Stigma

Style

Petal

Sepals

Ovary

Nectar Area

Nectar

Ovule/Seed

This illustrates the parts of a flower, showing an anther dehiscing (shedding) pollen, and the rest of the floral parts necessary for reproduction.

seasons depend on how close you are to the equator, no matter which side of it you live on. The closer you are, the more the seasons blend; the further away you are, the more distinct they become. At the equator, the climate is tropical. As you move farther away it becomes semitropical, then moderate, and finally polar. In the northern hemisphere, spring begins on about March 21 and summer about June 21. Conversely, in the southern hemisphere, spring begins on about September 21, and summer starts about December 21.

Equally important is having a general idea of what the plant world is doing during each of these seasons. Knowing and managing colony activities in normal, rainy, or dry seasons, nectar or pollen dearth, and regular seasons, gives you a head start on bee management. Just as with your garden, you should learn when to plant, when to water, and when to expect to harvest.

This then is the macro-environment in which you live. But closer to home, where your bees go, is a microenvironment that's equally important. Specifically, what plants are growing close enough that your bees can visit them? When do they bloom? The climate and the weather will affect, to a degree, when those plants bloom—let's say apple blossoms—by a couple of days, or as much as a week earlier or later in some years. All plant growth is dependent on day length, temperature, and available water and nutrients. In beekeeping literature there are a multitude of resources available that review those plants that are nectar and pollen producers and when they bloom throughout their growing range.

Nasonov Pheromone

When working a colony, you will often see worker bees on the landing board, facing the entrance. Look closely and you'll notice that their abdomens are raised in the air with the tip bent down just a bit. This position separates the last two abdominal segments, exposing a bit of the whitish integument below. Located at that spot is the Nasonov gland, which produces Nasonov pheromone. Exposing that gland allows some of the pheromone to waft away. To help distribute this sweet-smelling chemical, the bee will rapidly beat her wings. She is said to be fanning or scenting. This is intriguing behavior, and only workers can do it. Broadly speaking, this is an orientation signal produced to guide disoriented, lost, or following workers back to the hive.

Interestingly, when one bee begins fanning, it stimulates nearby bees to do likewise, and those bees that return begin to fan also. Very quickly you'll see many, many fanning bees on the landing board or the top edge of an open super, guiding their lost nest mates home.

This pheromone is also part of the glue that keeps a swarm together and all going in the same direction as it leaves its nest when heading for a new home.

Workers use this pheromone in a variety of other ways inside and outside the hive. You may see bees scenting at a source of fresh water. What you'll notice most, though, is that when you open a colony, the natural upward ventilation, partially driven by the body heat of thousands of bees, wafts up the commingled, subtle aromas of curing honey, stored pollen, and a good bit of Nasonov pheromone. This cocktail produces the distinct smell of a beehive. This is what makes all colonies smell mostly alike, but all a bit different. There is no other aroma quite like it or quite as attractive, to both a honey bee and a beekeeper.

When a hive cover is removed, some bees will fly away. If they are unskilled flyers or too young to have flown orientation flights, they may become lost, almost immediately, because they don't know landmarks. Skilled or experienced bees that fly off will immediately return and begin scenting behavior by exposing their Nasonov glands and fanning their wings to drive the pheromone away from the colony. The inexperienced bees will pick up on the pheromone's aroma and follow it home.

This swarm has left its colony but has not yet found a new home. You can see bees scenting with their Nasonov glands exposed. Other bees are dancing, trying to convince the swarm that they have located a good home site.

Review and Preparation

A brief review of what has been presented would be beneficial. So I will go over getting started, the equipment you'll need, the bees, and how to handle them throughout the season.

Before You Begin

> First thing, find out the rules and regulations that exist about bees and beekeeping where you live, or where you are going to keep your bees if not in your backyard.

> Check out all this with your family. Don't assume everybody is going to be as excited about bees in the backyard as you are, even if gardening and growing fruit trees are already a way of life. Be realistic about what you need in terms of yard space required, flight paths, equipment storage (and construction, if you go that route), cost, harvesting, and even beeswax activities if you decide to make candles, creams, or lotions. These will require room, time, and resources.

> Make sure you and your family members aren't in that tiny minority of the population that has honey bee allergies. But don't confuse normal bee stings with troublesome, but far less likely, allergic reactions. If you are unsure, get everybody checked out by a physician so that there is no doubt. Be safe.

> Next, find out what your neighbors think of your new-found hobby. Even if there are no, or only limited, restrictions on having bees in your yard, if a neighbor has an irrational fear of honey bees (and certainly if they have a health issue), your plans may evaporate. You can pursue this, but living next to a hostile neighbor makes life difficult for everyone.

> Be realistic about your available time when deciding how many colonies to manage. Even if local regulations allow up to five colonies on a lot the size of yours, do you have the time to take good care of that many? Begin with fewer hives, so that you can learn the ropes at a gentle pace.

> Learn as much as you can. Read books and magazines. Join the local beekeeper's association. Visit other beekeepers, buy or rent videos so that you see how to do the things you'll need to do, not just read about them. Take a beginner's class. Get all the catalogs.

Getting Ready

> Set your calendar so that your bees arrive right about the time dandelions and fruit trees bloom where you live. The following preparations have to be done by then.

> Prepare the place the colony or colonies will be. Provide screens so that the colonies aren't visible to neighbors, or from the sidewalk. Make sure flight paths are up and away from where people spend time—decks, the garden, play areas and especially neighbors' yards.

> Make sure your hive stands are tall enough to be out of reach of animal pests, and sturdy enough to hold one or more colonies weighing up to 150 lbs (70 kg) each, by season's end, and that the area around them is cleared and weed free.

> Provide a permanent source of water that will not dry up if a natural source is not available.

> Order your bees from contacts made at meetings or classes, or from suppliers, 4 or 5 months in advance, if possible, to ensure availability and timely delivery. Ask local beekeepers what race of bee works best where you live, and why they like them. Overwintering, honey production, pest resistance and gentleness are all traits to consider.

Equipment Dos and Dont's

> Don't for one moment be swayed by manufacturers' beginner's kits. These kits use the least expensive, poorest quality equipment, tools and protective gear that they sell. Customize *your* operation so that it fits you—not what somebody else chooses for you.

> Order your equipment early enough that you have ample time to prepare it all. Find a supplier that you are comfortable with, one that has all of the equipment you need, and is willing to answer your questions. When purchasing additional pieces, remember that not all manufacturers' equipment matches— stick with a single supplier, at least until you are more familiar with beekeeping.

> For convenience, safety, and ease of use, I recommend starting with 8 frame, preassembled equipment, using screened bottom boards with tray inserts, and wooden frames with plastic, beeswax-coated foundations. Either flat or decorative covers are effective, but landing board stands are not. Use queen excluders, and feed with one-gallon plastic pails.

> Paint or stain the exterior-only parts of your equipment with neutral or natural colors. White stands out, brown and gray don't.

> Get a full bee suit (best) or jacket-style suit with attached, zippered veil. Start with 2 hive tools, the large-sized stainless-steel smoker with heat shield and the least bulky, most sensitive gloves you can find.

> Decide what crop you will seek—liquid honey, round comb sections, Bee-o-pacs or perhaps none at all. Purchase accordingly.

Beekeeping Preparations

> Keep good records, and refer to them before examining your bees.

> Have a plan when opening a colony—know what you want to do, what you should be finding, and what maladies, if any, may be present.

> Become familiar with how a colony of bees operates. While learning, examine your colonies often, checking the broodnest boxes for signs of a healthy, productive queen, healthy brood and the right amount of it, plenty of pollen and honey stores, the right mix of workers and drones for the time of year, and any maladies that are present.

> Manage your colony for swarm prevention and control, but know that some colonies will swarm, no matter what you do. Have a bait hive in your yard, if possible. Answer swarm calls with care and caution.

> Requeen every colony every year with queens from producers actively seeking genetic resistance to pests and diseases in their bees. This is the only long-term solution to reducing pesticide use in your hives. Ask hard questions, demand good answers, and expect to pay top dollar. Recall that bad queens

never get better, and cheap queens are just that.

> Be patient when introducing new queens. Leave the cage in the colony for 3 days before removing the cork, and then allow 3 more days for the bees to remove the candy. It's the safest way there is.

> Monitor pest populations all season and be prepared to act if they suddenly increase. Treat if required, but don't treat "just because."

> Actively follow integrated pest management (IPM) techniques and methods to avoid or reduce pest populations and hard chemical controls.

> Always remember that honey is a food, and your family will eat it. Treat it as such.

Before you continue to the next section, here are a few guidelines to keep in mind. They're not carved in stone, but maybe they should be.

> There isn't a lot of work to keeping bees, but what you have to do, you have to do on time.

> Always dress and work with your bees so that you feel safe and secure, and your bees are not threatening anyone nearby.

> Your family, neighbors, and friends should not have to change their lives because you did (unless they want to).

> You should care for your bees the best you can, recognizing your responsibility for their well being.

> Beekeeping should not overwhelm your life. It is part of your life, not all of it.

> Sometimes you will be hot while working with bees.

> When it's not fun anymore, it's time to hang up the suit. If that's the case, plan an exit strategy by selling or otherwise disposing of your colonies and the bees. Don't abandon them, leaving the equipment to become a source of disease or pests and an attractive nuisance. Be a good neighbor.

CHAPTER 3 →About Beekeeping←

Introduction

This chapter is going to work from the ground up, starting like you do—with a brand-new package in your brand-new equipment. We'll shepherd that colony through its first year and into the next season, so next year you'll have all the information you need to continue.

Start by reading the section on installing a package. Become familiar with the sequence of events before you begin. Installing a package is a simple process and difficult to bungle. However, if this is your first time, it can be a bit nerve-racking.

Lighting Your Smoker

Your smoker, hive tool, and protective gear are all tied for first place as need-to-have beekeeping tools. But you also have to light and keep your smoker lit. If it goes out when you take your colony apart, you have a problem, and you'll need to retreat to relight. Meanwhile, your colony stands open and unsecured.

Lighting a smoker is similar to starting a campfire or a fire in a fireplace. Start with rapidly burning, easily combustible tinder. Newspaper is perfect and readily available. Once it is burning, add less-flammable material and establish that. Then, if needed, finish with longer-burning fuel that will sustain the fire for as long as needed. The following nine-step process works every time:

Step 2: Crumple the newspaper into a loose ball that doesn't quite fit into the smoker and light the bottom.

Step 3: Let the paper catch fire and the flame begin to move up without it reaching your hand. Push the paper to the bottom of the smoker using your hive tool, and puff the bellows gently two or three times to keep fresh air moving past the burning paper. It will flame up to or just over the edge of the top of the chamber.

Step 1: Assemble your tools: paper, pine needles, punk wood, matches, and smoker.

Step 4: Puff two or three times more until most of the paper is burning—but not yet nearly consumed—and add a small handful of pine needles to the top.

Step 5: Puff a couple more times until the flames from the paper reach up and catch the needles afire. Once they are burning well, push the burning needles down into the chamber with your hive tool, puffing slowly so air moves through the system.

Step 7: When the second, or perhaps third, batch of needles begins to smolder from the bottom, add the more durable fuel, if needed, on top of the needles. Keep puffing slowly to keep air in the system.

Step 8: When lots of smoke rises when you puff, and if the fire doesn't quit when you don't puff for a minute or so, close the smoker, still puffing occasionally.

Step 6: When the first small batch of needles begins to flame up, add more pine needles, loosely. Puff several times so you don't smother the fire. Keep air moving through the system. This is when most lighting attempts fail because not enough air is coming up from the bottom and none is available from the top. The smoldering needles starve for oxygen and the fire dies.

Step 9: When lit, the smoker should smolder unattended for many minutes. If it sits idle for a while without use, puff rapidly a couple of times so the smoldering coals flare up a bit, producing lots of cool, white smoke to waft over the bees.

When you finish using the smoker, empty its contents on a fireproof surface, making sure the coals are out. You can also lay the smoker on its side; without a draft, and it will quickly go out.

Still lit or still very hot smokers can cause fires if placed too close to flammable material. Make sure it's empty *and* cool before putting it away.

Occasionally check the intake air tube on the bottom to make sure it is clear, and scrape out the inside of the funnel. Accumulated ash will slowly close the hole.

If sparks from the fire are coming out of the spout, check your fuel because it may be nearly gone. If the fuel is still plentiful, grab a handful of grass and put it on the top of the fuel inside to stop the sparks.

Never aim your smoker at someone and puff. Besides causing the inhalation of smoke and limiting sight, flying sparks may ignite clothing. Also, use smoke sparingly on your bees. A little bit goes a long, long way in controlling the bees.

Installing a Package
When They Arrive

Almost all beekeepers get their first bees from a business that sells packages. Bees are sold by the pound, and the most common size has 3 pounds (1.4 kg) of bees. With about 3,500 bees in a pound, a 3-pound (1.4 kg) package (the most common and most recommended) contains more than 10,000 bees. Almost all are workers who have been removed from a single colony and placed in the package. Then, a queen, in her cage, is added to the package, a feeder can put in its slot, and the top covered.

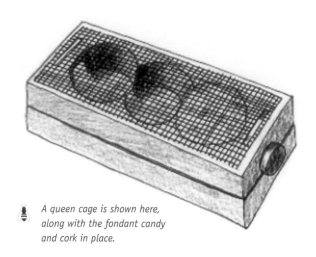

A queen cage is shown here, along with the fondant candy and cork in place.

You'll get your package from a local supplier or through the mail. Order it far enough in advance to assure there are packages available for delivery when you want them. A good rule of thumb is to have your bees arrive when deciduous fruit trees or dandelions are about to bloom—April or May in the northern hemisphere, October or November in the southern hemisphere. Therefore, your equipment should be ready before the bees arrive.

If possible, when ordering your bees by mail, make arrangements with the supplier to have them arrive midweek. Weekend mailings can sometimes be neglected in large post offices. Make sure your phone number (home, work, or cell phone for daytime messages) is prominently displayed on top of the package, with instructions to call you when the package arrives at your local

Shown is a typical 3-pound (1.4 kg) package of honey bees. The bees hang around the queen, who is suspended inside the package next to the tin can that holds sugar syrup. The package has a cardboard or plywood cover on top, keeping the can, the queen, and the bees in place. A few dead bees will be on the bottom of the cage; if there is a ½" (1.3 cm)-thick layer, contact the supplier.

Local suppliers drive to the package producer and pick up packages directly, reducing the delivery time and the stress on the bees. You can get packages either with a queen or without one and install the one you want later. Bees are a commodity, but the queen is the future of your colony. Don't settle for just any queen.

post office. Remember to inform the post office that you must retrieve the package immediately. Some post offices will store it in an out-of-the-way place, such as a loading dock, or storage room. Unfortunately, these places can be too warm, which can be lethal to the bees if held there too long.

If you purchased the bees from a local supplier, you'll pick them up on the assigned day—usually a weekend—right after they arrive.

In either case, get the bees home as soon as possible. These bees will be stressed—they have been several days away from their home colony with limited food, a strange queen, temperature extremes, and jostling.

Of course, you should get the bees into their new home as soon as you can. That may be just an hour or two, or a day or two, depending on your schedule and the weather. Cool, rainy weather is tolerable, but very cold temperatures—below 40° F (5° C)—are probably colder than you want, so wait a day.

While waiting, keep your package in a cool, dark place, such as the basement or garage. Put down a couple sheets of newspaper to keep the floor clean.

You'll also need to feed the bees. Get a new, unused spray bottle (such as those used for misting plants), and mix up a sugar-water solution of one-third sugar to two-thirds water by volume, that is, one cup white sugar mixed with two cups warm water. As soon as possible, spray some of this solution through the screen directly onto the bees. Don't saturate them, but moisten as many bees as possible through both sides of the package. Do this at least a couple of times a day as long as they are in storage. The feeder could be plugged or empty, so don't assume they have food available.

In They Go!

You've prepared your family and your neighbors for your adventure long before this day. The site has been chosen and prepared, and your equipment has arrived and is ready.

If time works in your favor, plan to install your package in late afternoon or even early evening. This timing helps your bees remain calm and assists in settling in for the evening.

To begin, use your spray bottle to give the package a feeding. Then, take all your equipment out to the site and get it set up. Put the bottom board on your hive stand and one super with frames intact on top of that. Leave the next two supers, without frames, stacked alongside, and one super with frames intact. (You need 4 supers.) Don't forget the inner and outer covers. Prepare the sugar syrup for the feeder pail with a two-to-one water-and-sugar solution. Fill the pail to the very top, and make sure the lid is secure. Refill your spray bottle at the same time.

If all your equipment is new, your frames will have only foundation, either wax or wax-coated plastic. If you have combs from other equipment, place those in the center of the eight frames in the box. (Be certain, however, that the colony from which these frames came was disease free.)

Use a mister bottle to put sugar syrup on the screen so the bees have food. Use a new bottle, and don't soak the bees when misting.

The only tools you'll need are pliers, your hive tool, and the mister bottle.

Put on your protective gear (for more on protective gear, see page 22), light your smoker (you probably won't need this, but be prepared), make sure you have your hive tool and pliers, and bring your package to the hive site. If you have one, bring a board large enough to fit securely on your hive stand to serve as a solid, dry working surface.

Remove the cover, inner cover, and six frames from your eight-frame box. Set the cover, inner cover, and frames behind the box or to the side. If you are installing more than one package, prepare all the boxes at the same time.

Set the board next to your hive on the stand, if there's room, or on the ground next to it.

Position yourself behind the colony into which you plan to install the bees. Make sure you have all your tools—smoker, hive tool, and feeder. Set the package on the board, and using your hive tool, remove the cover. You may also need pliers for this. Remove any protruding nails or staples, and keep the cover close at hand.

🐝 1. Make sure you have a good hold on the can before thumping the cage. If not, it may fall off the support below, if there is one.

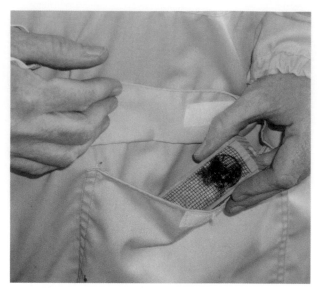

🐝 3. When the queen's cage is free of worker bees, put it in your pocket to keep her warm and for safe keeping.

🐝 2. Remove the queen's cage carefully. Don't drop her, if you can help it. Don't let the bees unnerve you. Gently shake or blow them off the cage.

🐝 4. Quickly and carefully dump the bees into the space created by the missing frames. Shake the bees a bit to get them to come out, but don't worry about every last bee. You'll get them later.

Take a look at the opening underneath. You'll see the top of the feeder can, flush with the surface of the package top. There will also be a slot cut in the top with a metal strip in it. This strip holds the queen's cage, which is suspended below in the mass of bees. It also allows you to easily remove the queen's cage. The strip should be long enough to grasp easily and to hook over the top of a frame.

Lightly mist the bees again. Using the corner of your hive tool, lift up the feeder can. Some are easy to catch and lift, but some will be a bit below the surface and more difficult to grasp. If you simply can't catch it, try lifting just a bit and grasping with the pliers. Hold it with one hand, and grab with your other hand.

Next, lift the package and thump it on the board so all the bees clinging to the feed can let go and fall to the bottom. Don't worry—they're covered in sugar syrup and don't care one bit. You may want to lift the can 1" (2.5 cm) or so and hang on to it while bouncing the package. Try lifting and moving the package to make sure you have a secure grip. When comfortable, lift the package 1' (30.5 cm) or so and thump. Lift out the can, set it on the board, and slide out the queen's cage—without dropping her. Cover the hole in the package with the cover. Check to make sure the queen is alive and moving, and then put her cage into your pocket to make sure she stays warm.

Check to make sure the frames have been removed and that you have placed the wooden entrance reducer in the box into which you are going to dump the bees. Make sure the box that is going to go on top of the bottom box is close by, along with the inner cover that will go on the new colony.

When you're ready, you're going to thump the package again, then remove the cover and pour the bees into the cavity in the box created when you removed frames. Pour as many as you can—shake them a little, but not much—and set the package, still containing a few bees, in front of the colony. This action will put some bees in the air, but don't be concerned. They are home-less, very confused, and not likely to cause you harm.

Carefully lower the frames you removed back into the bottom box. Let them slowly sink as the bees on the bottom are gently

5. This is what it will look like when the queen's cage is placed correctly between the bottom bars of the frames in the top box. Make sure the screen side is facing down and is unobstructed so the bees can feed and make contact with her. This placement is very important.

6. Place the feeder pail over the hole in the inner cover. You'll need two medium supers to enclose the pail.

OOPS!

"I dropped the queen's cage! Help!" This happens, especially when you're wearing gloves. You have to get it. Here's how: Quickly put the cover back on the package. Look for the queen cage in the mass of bees inside. If you don't see her, gently roll the package until you spot the cage. If you don't already have a glove on, this may be a good time to get it. Gently thump the cage to confuse the bees and spray them with syrup. Remove the cover, and reach in, grab the cage, and pull it out. Quickly replace the cover. It's that simple. If you can, don't wear your glove. The texture and feel of all those nonthreatening bees on your fingers and hand is a feeling like no other.

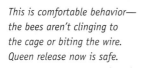
Remove the cork on the candy end of the cage, using the corner of your hive tool. Don't remove the cork on the other end or the queen will walk out prematurely.

pushed out of the way. After you have replaced all the frames in the box, put on the inner cover for a moment. Grab the second box with frames that you brought with you and stand it on one long side. Take the queen's cage and place it between the bottom bars of frames, just off center so if your feeder leaks, the flood won't drown your queen.

Holding the frames secure, remove the inner cover and place this second box on top of the box with all the worker bees. Replace the inner cover, position your feeder pail on top of the inner cover so that the screen is centered over the hole, add two supers to cover the feeder, and then place the top on.

And you're done. Make sure the small entrance to the reducer is open to let the remaining bees in the cage go in at their leisure (usually overnight is fine). Pick up all the extra stuff and tools you brought to the site and call it a day.

If you're still uncertain, reread this section so the sequence is clear in your mind. It's fairly simple and straightforward, but being familiar with it always helps.

First Inspections

You will run across a variety of recommendations on how often to inspect your colony, especially a brand-new package. Checking the amount of available sugar syrup doesn't count here because checking the pail feeder doesn't intrude on the colony. The recommendation made here regarding releasing the queen, however, runs counter to the conventional recommendations. Specifically, place the queen in the colony during installation, but do not allow her to be immediately released. Rather, keep the entrance blocked with candy covered with the cork or tape for three days. This double-visit technique increases queen acceptance significantly, and reduces this stress on the colony during a time of many stresses.

After the package has been installed, your tasks are over for a time. Wait patiently. Then, in three days (or four if the weather isn't warm and sunny), inspect the colony. This delay allows the queen to be released safely from her shipping cage. Look carefully at the cage when you first separate the boxes and *before* you smoke the bees. Are workers clinging to the cage, trying to bite the screen, and refusing to move, even when smoked a bit? If this is the case, put your colony back together and wait a couple of days. This is good advice if you look and just can't tell. Err on the side of caution and be patient.

However, if the behavior seems friendly, proceed slowly. Separate the two supers and lift the top one with one hand far enough

This is comfortable behavior—the bees aren't clinging to the cage or biting the wire. Queen release now is safe.

Queen Replacement

When you lose a queen during introduction, immediately contact the business from which you purchased her and let them know what happened. They may offer a free replacement, but you pick up the shipping. Don't argue or debate their decision. They may have good advice on introduction techniques, or perhaps they have had mating problems and a replacement is due. Or you may have goofed. However, if they become defensive or obstinate (a rare occurrence), consider a different queen producer.

When the queen arrives, introduce her using the same techniques as before, mindful that this new queen didn't have two- to three-day acquaintance time in the package. After 10 or 12 days, you need to inspect the broodnest again to record the quality and quantity of the brood the queen is producing. There should be eggs, open brood, and sealed brood roughly in the proportion of 1:2:4. That is, there should be twice as much open brood as there are eggs, and nearly twice as much sealed brood as open brood. This may vary widely, depending on how long the queen has been laying, so put some—though not too much—stock in this. Later, the ratio is important.

Also look at the sealed brood, noting solid patterns in the center of the frame surrounded by open brood and eggs. Drone brood should not be present anywhere in the center. A drone-laying-queen needs to be replaced immediately, as does a queen who has a lot of empty cells in her broodnest.

As difficult as it may seem, and as frustrating as it can be, never debate replacing a queen who isn't performing well. Bad queens never get better, and a bad queen will, at best, head only a mediocre colony and may cause its demise and death. Requeen at the first signs of a problem. It's a small investment in an entire season. Don't be cheap when it comes to queens—you will always be sorry.

Something to keep in mind about your package: No new bees were produced for at least 21 days. Usually, it is closer to four weeks. During that time the bees that came in your package have been making beeswax comb, tending the queen, feeding the young, foraging, and guarding.

This three- to four-week period is about as stressful as it can get in a colony. The demand for food, especially pollen for the house bees to turn into worker jelly, is extreme. Sugar syrup is a wonderful carbohydrate source, but protein is needed also. Foragers are scrambling to collect pollen, especially because you are providing sugar. It's a hard time in your colony until the new bees begin to emerge.

to expose the queen's cage. Lightly smoke the bees so they retreat. By now they are generally quite friendly and ready to get going. Retrieve the cage and gently lower the box, trying to avoid squashing bees on the edges. Remove the cork from her cage using the corner of your hive tool to expose the candy. Poke the candy with a twig or small nail to ensure the candy hasn't dried. If it has, poke a hole all the way through it with a small-diameter nail or piece of wire, being careful not to impale the queen on the other side. If you're having trouble doing this outside, put the cage in your pocket and take it indoors to do the job.

When complete, replace the cage in the same place, making sure the screen isn't covered by any new wax comb that may have been built up around the cage or frame parts. In fact, you'll often find stray slabs of beeswax comb, so carefully remove them with your fingers or hive tool first. Replace the box carefully, check the feeder, refill it if necessary, and put everything back together.

The queen should be released by the bees in another three days, so plan on checking again. If she's not, and if the behavior of the workers is still protective rather than aggressive, pull back the screen and let her walk out, heading down between frames. Don't let her fly away. Close up the colony.

After her release, the queen should begin laying eggs in a few days, at least within a week. After that time, check for eggs.

You'll find them in the center two or three frames, probably near the tops of the frames in the bottom box. Look at all the frames that have some comb built on them if you don't find eggs in the center. Recall that eggs are tiny, standing straight up on end, and nearly the same color as the new wax—look carefully.

Once the queen is laying eggs, you're over the first hurdle and don't need to inspect the broodnest for another 10 to 12 days. Make certain the feeder stays full, however. The bees will continue to use the sugar syrup for some time, especially when the weather doesn't cooperate—a sure thing in the spring—and at night. For much of the first season, the colony is living hand-to-mouth with little chance to build reserves. The more you can help, the better off they will be.

So, what if you don't find eggs? It happens. Perhaps the queen wasn't mated at the producer's before she was sent, or the disease nosema affected her, or she was injured in transit or installation, or in spite of your best efforts and judgment she wasn't accepted by the colony and was killed? Look to see if she's there. In a colony this small, if she is alone and walking around she will be fairly easy to find. Listen for the telltale buzzing sounds of a queenless colony.

If you find her, close up the colony and give her two more days to start laying. If after all this nothing's happening, something's wrong and she needs to be replaced. If you don't find her, order a replacement just as soon as you can—that day, if possible.

Honey Flow Time

A newly introduced queen starts laying eggs at a slow pace. She has already laid some eggs before the queen producer sent her to you, but then went for many days confined to her cage. Once released, this young queen's egg-laying rate builds slowly, starting at perhaps 100 eggs a day for a bit, gradually increasing to as many as 1,500 per day when all conditions are favorable. The rate depends on the health of the colony, available pollen, favorable foraging weather, and adequate space.

The bees generally begin building comb in the center frames, using most of the space from top to bottom in the bottom box and anywhere from none to all of the frames from top to bottom in the second box from the bottom. Usually, it's most of both, but the bottom of the frames in the top box and the top of the frames in the bottom box get built first. The ends of the frames also are left to fill later.

When there is comb being built on most of five or six frames in the bottom box, and four or five in the top box, it's time to add a third box. At the same time, switch positions of empty frames and frames with comb, but not brood, placing those with some comb next to the edge of the box. This rearrangement encourages the bees to fill all combs, rather than use only the center frames in the boxes. If you don't switch frames, the bees may "chimney" their living quarters, building all the way to the top but only in the center. During this buildup time there may be a nectar flow from early and midspring blooms. Nonetheless, maintain the feeder at all times to ensure there is always sugar and water in the hive.

Here's a handy tip: When you add the third box for your brood, add the queen excluder and check and make sure you have two or three honey supers ready to add. If not, get them ready. Another tip: Paint the supers different colors so you don't inadvertently use a honey super for a brood super or vice versa.

As before, purchase preassembled boxes and wooden frames with plastic foundation. Get them painted (or stained) right away and have them ready. It takes an hour or less for two or three supers. If you have to assemble them, plan on a lot more time.

After two to three weeks, depending on the weather, the third box should have three or four frames with comb and two or three with some brood. The broodnest should encompass one or two frames in the top box and most of the frames in the middle and lower box, with pollen and honey stored on the edges, and mostly or all honey on the very edge. The outer side of the frames on the edge may be barely filled with comb and a little honey this early in the season. Again, exchange frame locations to encourage the bees to fill the partial frames in all boxes.

Be patient—it may take twice as long as described to set up your colony. The key is to observe the sequence of events and the buildup of the whole colony. Weather, available protein, the building of new comb, and you, the beekeeper, all add stress to this new colony.

How Many Supers Should You Have?

If you are in a warmer climate, the nectar flow tends to be early, intense, and short, maybe only three to four weeks. Sometimes, it's less. Supers can fill fast, and room for incoming nectar can disappear almost overnight. Without room for nectar or honey, the rate of foraging will taper off. It may even stop. Or, the broodnest area may begin to fill up, reducing the queen's activities. For short, intense honey flows, put two supers on right away. This makes sure the queen isn't slowed and there's room to grow, especially if you miss a week's monitoring while on vacation or if you get busy in the garden. If the flow stops, the bees won't use the box and you can remove it unused.

There are mixed reviews on the best way to manage this situation, but if you restrict storage space and there is lots of incoming nectar, the bees will (almost always) begin to fill the broodnest, restricting egg laying. Recall the conditions associated with swarming behavior—restricted space, lots of bees in the hive, a queen whose egg laying has slowed, and a whole world to explore.

This is where choosing medium-depth, eight-frame equipment is truly a benefit. When it's time to add that additional super, first remove the partially filled box, then put the new box directly on top of the queen excluder that was added when the third super was put on. Your bees, accustomed to the excluder, easily move up to store nectar and honey, will continue to do so, filling the top super. On the way up they pass right through the new space and will begin to take advantage of it, *as long as nectar continues to come into the colony*.

Adding a honey super provides the room necessary for colony expansion. But don't be caught short. Late spring nectar flows can be intense, and if your colony's population is strong, the right plants in bloom, and the weather suddenly turns warm and humid, that super can be filled in a week or ten days. It is astonishing how fast this can happen when everything works. Being slow will reduce the honey crop at the end of the season, but that's not a negative if you're not sure what to do with hundreds of pounds of honey.

More likely though, nectar flows are slower, and the bees will take substantially longer to fill these supers, mostly due to erratic weather. Another factor that can be limiting is available forage. If your colony is near undeveloped land where a diversity of blooming plants exist, the nectar flow will speed and slow as nectar-producing plants flourish and fade. In a more developed area, nectar plants are primarily domestic and the variety can be nearly infinite, but the quantity can be limited. You need to check on the rate of filling in your location, and add more space either as needed or as you want to plan your crop size.

When the first honey super has comb on five or six frames and honey being stored in four or five, even if it's not yet covered with wax, it's time to add a second super for additional space for nectar storage and, eventually, honey storage.

Keeping Records

If it hasn't occurred to you already, keeping a log of your colony's activities and progress is a good idea. Particularly for the first couple of seasons, making notes will force you to attend to the fundamentals, and the notes remain as a record of what happened when. If you have several colonies, some notes will serve for all, but each colony, you will learn, has its own distinct personality, requiring similar but different management actions. At the beginning of the season, record the following for each colony:

> Queen source and/or package source
> Location, position (north-, south-, east-, or west-facing) of the colony, and registration papers and inspection reports if you have them
> Condition of equipment

Then, include the following notes about each visit:
> Weather and time of day
> Date
> What's blooming—you need to look around a bit for this, but the major common plants are easy: dandelions, wildflowers, flowering shrubs in your yard. These major nectar-producing plants bloom about the same time each year, but you need to learn when so you can prepare for honey flow time.
> Temperament of the colony—easy, busy, flighty, fussy, loud

> Depending on the time of year, note:
>> Queen cups present
>> Queen cells present
>> Spotty brood pattern
>> Number of eggs, open and sealed brood (1:2:4 ratio)
>> Drones present
>> Honey and pollen present
>> Signs of pests or disease
>> Physical condition of combs and other equipment

Then note what activities you performed, such as:
> Requeened (note the source and breed) and color of mark—always get marked queens.
> Fed, and how much
> Exchanged frame positions
> Removed old, black frames and added new frames (and where you got these, and what did they cost? It's not a bad idea to write on new equipment where and when you got it so you remember particularly bad or good suppliers.)
> Added brood or honey supers
> Couldn't find queen, or found a queen that shouldn't be there (your marked queen is gone)
> Applied medications (and when the next treatment is due)
> Harvested, and how much

Reviewing your notes before you next open your colony will remind you of actions you need to take, equipment you need to buy, problems you need to check, and what you should expect to find.

Keep records in a large notebook that is difficult to misplace. After a few seasons, your notes will be minimal because you will have mastered the routine management skills, noting only swarming dates, new queens added, medication applications if used, and the amount of harvest.

Opening a Colony

The first few times you examine a colony can be exciting, scary, and confusing. It's easy to get sidetracked and forget to do one of your planned tasks. So before you get started, make a good mental note of why you are looking inside. This is good advice whenever you are going to examine your colony. Check your record book first, always. When you know why you're going in, you'll know what you need to do the job, so get everything together and bring it with you.

Before you begin, make doubly certain your smoker is burning well, but give it a reassuring puff every few minutes, just to be sure. If you're feeding, have your extra feeder full and ready. Bring supers or other equipment you may be adding, and have your hive tool in hand. It's not good to leave a colony open if you have to run back to the house.

This may seem obvious, but quickly scan the ground for rocks, branches, or toys. Stepping on one of these with your hands full of a box of bees can be disconcerting. Check to make sure nothing is on the hive stand where you plan to set your boxes, either. And don't put the things you brought with you in a place where they will be in the way before you use them. Whenever possible, minimize moving things more than you need to—it saves the back and the temper.

When everything's assembled, you're ready to puff the tiniest puff of smoke into the front door. This will offset the guards at the front immediately, reducing any flying. Then, step back and again look at the setting, making sure nothing's in the way and you have everything you need. This delay of a minute or so lets that little bit of smoke waft up into the lower box just a bit and contact the dance floor area, slowing that behavior also.

Remove the cover as carefully and as quietly as you can, but keep it close. If you have a telescoping cover, place it top-side down next to the colony on the hive stand, about 6" to 8" (15.2 to 20.3 cm) from the colony, with the long side parallel to bottom board. When you remove the inner cover, place it on the cover, turned 90 degrees. Place any boxes you remove on the inner cover with the same orientation. This placement keeps bees from leaving the box from the bottom, and a bit of smoke puffed over the top will keep them inside. If there are boxes protecting your feeder, remove them and the feeder, putting them out of the way. Puff a half puff or so of smoke into the center hole before you remove it. Again, wait a moment or so before proceeding.

🐝 *Pry up the inner cover.*

Pry up the inner cover with your hive tool, opening it a couple inches, and give another half puff or so of smoke. If bees are flying out, give a couple puffs. Slowly remove the inner cover and set it on the cover. Look down between the frames. What do you see? Between the middle three or four frames will probably be lots of bees and some comb built up. Perhaps nothing has been done in this box yet, if the package is still young. If there is nothing, lift up a corner of the box and puff two or three times, then slowly remove that box and put it on the inner cover.

There should be lots of bees and built-up comb on the center frames in the box. If you are going to look for eggs or perform any other broodnest inspections follow these points:

> Stand behind or to one side of the colony.
> Using your hive tool, loosen the frame closest to the edge of the box, or the one second closest. If there's a difference, choose the side that is least built up. Loosen both ends of the frame if they are stuck.

🐝 *Remove the cover.*

🐝 *Loosen the closest or next-closest frame first. Use the curved end of your hive tool for leveraging and loosening frames.*

- Puff some smoke if bees are coming up between frames and crawling on the top bars and your hands. Watch for lots of bees lining up at the top of the frames, watching you.
- When both ends of the frame are free, lift one end with your fingers and the other with the corner of your hive tool if there is lots of comb, or with your other hand if the frames are not stuck.
- Keep your hive tool in the palm of your hand and between your thumb and forefinger, held in place with your little finger and ring finger. This grasp leaves your thumb and forefinger of each hand to hang onto the frame when you lift it up slowly.
- If the frame is empty of comb, carefully set it down on end at the front door so any bees can easily find their way home.
- Loosen the next frame if it's stuck by inserting the curved end of your hive tool between the two frames and twisting it. The leverage this position gives you is amazing, and it will loosen almost any frame.
- When the frame is loose, remove it slowly, especially if there's comb on even one side of the frame. Lift the frame straight up until the bottom is clear. This reduces the chance of rolling bees (or the queen) between moving combs and crushing them.

Slowly lift the frame straight up, so you don't trap, roll, and kill bees between the comb surfaces.

- Hold the frame by the lugs or shoulders between your thumb and forefinger.
- If there's no comb, lean the frame against the first frame, by the front door.
- If there is comb, you'll want to see if eggs or brood are present. Turn so the sunlight is coming over your shoulder. Hold the frame at midchest height and away from your body a bit. Tilt it

so the light shines directly down to the bottoms of the cells. Sunlight is the best light there is for seeing eggs or brood. If at all possible, keep the frames over the open box or the first box you removed so that if by chance the queen is there, she won't be lost if she falls off the frame.

Let the sun shine from behind you, directly over your shoulder, so that it shines down into the bottom of the cells.

- If you are going to look at additional frames, slide this one over to fill the empty space left by the two empty frames, keeping everybody inside and safe.
- Puff more smoke if bees are rising between frames or are starting to fly.
- After examining the next frame as before, slide it over out of the way and examine the next frame.
- When the examination is complete, carefully and slowly slide the frames back to their original position. Don't rush, and avoid squishing bees. When a frame's top is in place, look inside to make sure the bottom is straight down. If it is not, you will crush bees when the next frame comes over. Use your hive tool to push to straighten the frame, if necessary.
- When everything's back in place, quickly puff smoke on the edges to remove those bees.

When bees begin looking up at you from between frames, gently puff one or two small puffs of smoke so they go down and leave you to your tasks.

> Slide, rather than drop, the box back in place, pushing bees out of the way. You will probably squash a bee or two, but if you're slow and careful, you won't catch many.
> Replace the inner cover, feeder, empty box, and cover, and call it a day.
> As interesting and exciting as it is to watch bees go about their business when you're examining a colony, keep your visits to ten minutes or fewer depending on the day, of course. Cool, cloudy days require short work. Warm, sunny days stand a bit more time. At any rate, after a while, even the most tolerant colony begins to take a dim view of all this exposure and becomes less easy to work.

Honeycomb and Brood Combs

The frames you use in your boxes in which the bees have their broodnest are used much differently than the frames in which they store honey. When a larva is raised in a cell, the cell is filled with worker jelly for her to eat. Shortly before she pupates and spins her cocoon, she voids her digestive system into the bottom of the cell. When she emerges, the pile of waste (called *frass*) along with the remains of her cocoon stay behind. As soon as she leaves, house bees clean out as much of this as they can, but they

cannot remove it all. What remains is sealed with a thin layer of propolis and wax. As a result, after only two or three generations these cells begin to darken. After a few seasons, they will be nearly black. Added to this is all of the pollen, dirt, plant resins, and other material foragers bring back with them and spread over the dance floor area. Plus, spores from nosema, American foulbrood, and chalkbrood diseases are present in the broodnest, even if you are treating for these maladies. And any pesticides that may have made their way back to the hive are present on or in the wax.

All of this debris combines to produce a very dark comb, laden with things your brand-new bees don't need to be exposed to and that surely add a level of stress to your colony. Replace old, dark combs routinely to avoid this stress. Every three years, in the spring when most combs are empty, is a good recommendation, but it certainly should occur whenever the comb becomes so dark that when held up to the sun, no light passes through.

Mingling frames from the broodnest with frames from honey supers also causes problems. The material in the cells and the bees walking on this darkened wax will darken the honey stored in the combs and add bits and flavors of what was there before, reducing the pristine quality of the honey you want to harvest. The bottom line: don't mix frames used for honey with frames used for brood.

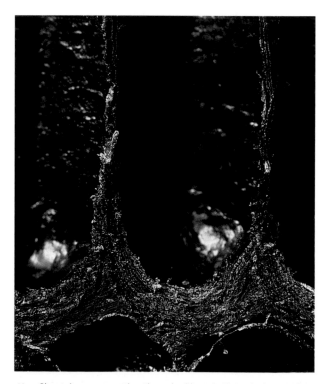

Shown is a cross section through old comb. Note the layers of cocoon, propolis, and general gunk. This comb should be replaced.

If your colony is progressing, the queen is working in all three boxes, and the workers are storing pollen and honey around the edges, it's time to add additional room—but not before at least six frames in the bottom two boxes have comb on *both* sides, and the same on at least four frames in the top. Don't rush the colony just because it *seems* to be doing well.

If spring has been late, or if good weather has come in spits and spurts, your colony won't build as fast as an established colony in the same place during the same time. Keep feeding your colony as long as they are taking any of the sugar syrup. They may take it for a few days when it's cool, and then stop when the weather warms, plants are blooming, and the bees can fly.

Sugar syrup may develop a black mold in the pail when the bees don't eat it for a couple of days, or it may actually ferment if the weather is warm. A good rule of thumb is, if the look or the smell of the syrup is such that *you* wouldn't drink it; don't give it to your bees. Sugar syrup is an inexpensive and easy way to ensure your bees don't run into the stress of a food shortage, even for a single day.

You don't want the queen wandering up into these brand-new honey supers to lay eggs and thus darken the honeycombs, so you need to provide a barrier. There are lots of management tricks you can try, but by far the easiest is to place a queen excluder between the top brood box and the honey super(s) above it. When the top brood super starts filling, move a frame (or two or three) with some honey (and no brood) into the super you are going to use for honey. What you are doing is laying some ground rules for management. Above the queen excluder—only honey. Below—honey, pollen, and brood.

The frames with even a little honey *above* the queen excluder send a message to the food-storer bees that this is an acceptable place for nectar and honey to be stored. Given all this room, and the open invitation to store the food, the food storers will (almost always) begin moving honey up, leaving the three broodnest boxes for mostly brood, pollen, and a little honey.

In the warmer regions, this activity is going strong by early spring (by early summer in the coldest areas), so be prepared ahead of time with the right equipment.

Once you've decided that your colony is growing at an acceptable rate (recall the brood ratios of 1:2:4), the queen is doing well, food is coming in at a pretty steady rate, the weather has calmed down, and available room for incoming nectar storage is running low (all frames have some comb and none are empty), you can consider adding additional supers for surplus honey storage.

At this point, the sugar syrup feeder has served its purpose and can be removed. If you don't remove it, some syrup may be taken by the bees and stored in the honey super. This isn't a critical mistake, but some of the honey you harvest will be sugar syrup, not floral nectar.

You *can't* have this happen if you are selling honey, but the bees don't care one bit. Remove the feeder, clean it well, and store it for future use. No matter where you are during the nectar flow, be mindful of the wrinkles that can occur. Check the *broodnest* every week or ten days, and the honey supers at the same time. The greatest stress on your brand-new colony the first season isn't going to be the common diseases and pests you will eventually encounter but, rather, the immediate challenge of establishment. Your goal the first year isn't a crop of honey, but a healthy, well-established colony. Next year, and for years after, your goal is management for production.

This doesn't mean, however, that you should ignore those problems that can occur, and now is a good time to learn the signs and symptoms because the colony is small and easily manipulated.

Medication Use and Integrated Pest Management

Your personal philosophy about using chemicals in your garden may range from using none to using just enough to control emergencies, to using whatever books, magazines, or cooperative extension bulletins recommend. When you keep bees, it is important to read pesticide labels carefully to make sure you use yard or garden pesticides or herbicides that are not toxic to bees.

There are also some important things to consider when you need to treat the bees for ailments. Like all animals, honey bees have their own set of maladies. Some are troublesome but not lethal; others can be deadly without some form of intervention. The actions needed to deal with pest and disease prevention, avoidance, maintenance, and treatment are integrated into the annual management schedule. Details about each pest and disease are set aside to point out weak spots in their cycles that you can take advantage of.

Each of these problems has a range of nonchemical management activities that can prevent or reduce their occurrence, keep their impact from becoming lethal, or reduce to the minimum the number of chemical applications needed.

There are two extremes in pest and disease management. One is removal from the scene—in a word, destruction. Compare this approach to discovering fatal diseases such as fire blight on apples or late blight on tomatoes—you must remove and destroy the infected plants or plant parts to remove any opportunity for further infection. With your bees, this method may mean destroying part or all of a hive, including bees and equipment. The other extreme is to prophylactically apply chemicals according to label instructions, so that pests and diseases are never allowed to flare up and are held in check for as long as the chemical treatments continue.

These two approaches are generally effective. But understanding the value of each of these techniques, and all the choices in between, should make your decisions easier and more compatible with your beekeeping philosophically.

Integrated pest-management systems employ a least-toxic approach by applying natural pest controls, such as predators or parasites, avoiding diseases by selecting good apiary sites, providing barriers to some pests, such as bear fences, or using strains of bees resistant to common problems. The last choice is applying chemicals; this approach is used only when the pest becomes overwhelming, all other choices have failed, and loss of the colony can be prevented only with a chemical application.

Maladies

Like all animals and plants, honey bees have a host of bacterial, fungal, viral, and animal pests that are looking for an easy lunch. Some feast on the honey bee's larval stage, some on the pupal stage, and others find the adults attractive. Most of these aren't lethal to a colony but can be to an individual. Others, when populous enough, can destroy an entire colony.

Most are seasonal, and you deal with them as a matter of course. Good management of bees is just like good management of tomatoes. You deal with hornworms, late blight, and aphids in their time, just like soil fertility, weed control, and harvesting.

Nosema

When buying replacement queens, you should know that one of the first questions to ask is how the producer deals with nosema in his operation. The same goes for the business that produces your package. Most, but not all, routinely provide an antibiotic to all of their colonies to minimize the chance for infection and especially to keep the queens healthy.

A queen infested with nosema will be superseded in about three weeks, period. If the producer doesn't treat the bees, or if the bees came from an unknown supplier, you're guessing about her health.

Nosema is a quiet killer, often compared to high blood pressure in humans. Symptoms are usually impossible to spot until an entire colony is breaking down, at which time treatment is useless. The disease is caused by a protozoan, *Nosema apis*, and infects the digestive system of its victims. It is spread from bee to bee by spores. It reduces the ability of bees to digest food, effectively competing with its host for nutrients. It shortens the life of an infected worker by about 10 percent and, thus, the amount of honey your colony could have produced. It also affects a worker's ability to produce brood food and causes her to age prematurely. If a queen is infected, her ability to digest and use food is reduced, which in turn affects egg laying. When this slowdown in egg laying is apparent to the workers they begin plans for replacement.

Bees that have been treated before you get them are usually without symptoms. However, this and a couple of other maladies are often referred to as stress diseases. That is, when bees are stressed for any reason, if the disease is present it could, and probably will, flare up, lying dormant or unobtrusive otherwise. You can certainly minimize the stress your new package experiences by feeding well (even offering a pollen substitute, if needed), examining your colony only when necessary, making sure the hive gets lots of sun for warmth, and has good ventilation and plenty of water.

Nosema spreads from colony to colony by drifting bees (especially during early spring) and by exchanging equipment between colonies. Infected bees confined in a colony during a long winter or dreary spring may eliminate waste inside a colony rather than outside, spreading the spores of the disease everywhere.

To reduce sources of infestation:
> Do not exchange frames or equipment between colonies.
> Replace combs contaminated with spores at least every three years.
> Keep stresses to a minimum, including other diseases and pests, poor ventilation, lack of food and water, and being queenless.
> Make certain colonies have good sunlight exposure in the spring.
> Reduce drifting between colonies, including robbing.
> Insist on purchasing packages and queens from producers who practice these techniques and treat their bees for the disease.
> Treat for the disease in the fall with the antibiotic Fumagillin, which does not cure the disease but does inhibit spore germination.

Chalkbrood

Chalkbrood, *Ascosphaera apis*, usually just called chalk, is a fungus, somewhat like powdery mildew on a plant leaf. It grows only in larva that are fed or come in contact with chalk spores in their cell. The spores germinate and end up in the gut of the larva, which starves in the competition for food. After the larva dies, the fungus covers the rest of it so the cell is completely filled with a mass of fungal growth.

When the fungus has essentially consumed the larva, it shrinks and is cleaned by workers. If conditions are ideal, the fungus will reproduce, spreading black spores everywhere. The most ideal environment is usually at the edge of the broodnest where temperatures fluctuate a bit.

You'll probably first notice these hard, white, or black-and-white mummies on the landing board or the ground in front of your colony. This is most notable in mid- to late spring, when the stress on a colony is the highest.

Package bees placed on new equipment usually don't experience this problem the first season or two, even considering the stress on

🐝 *You'll first see chalkbrood mummies on the ground, having been removed by house bees.*

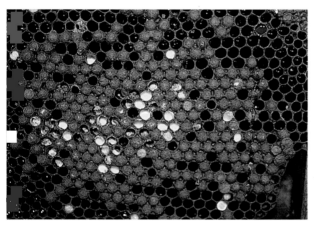

🐝 *You'll also see chalkbrood mummies in cells in the broodnest. They will be white, chalky, and hard. They may rattle when you move the frame. Some may have black spots on them, meaning the disease has matured to the point of producing spores, providing additional means of infestation.*

the bees and the likelihood of temperature fluctuations in the spring. There are no chemical treatments for this, but many management practices reduce the incidence and severity of chalkbrood.

In rare instances, the disease will overwhelm a colony, infesting perhaps a quarter or more of the brood. This happens when there are lots of available spores and cleaning behaviors don't keep up. To repair a colony like this, immediately replace the queen with one from hygienic stock, and remove as much of the chalk in combs as possible. You can destroy entire frames to accomplish this, replacing them with new frames. This helps reduce the source of infestation in a colony.

European Foulbrood

Another malady often associated with springtime stress is caused by a bacteria—*Melissococcus pluton*, commonly called European foulbrood, or EFB. It strikes very young larva, and is inadvertently passed on to them by nurse bees that spread it throughout the colony.

The larva consumes food infected by the bacteria, which inhabits its gut and then competes for food. The larva dies when still young, while in the bottom of the cell, and sort of melts down there, turning into a rubbery pile.

Your first clue is a brood frame with what appears to be random holes. It's often called a shotgun or spotty pattern. These scales, as the dead larva are called, are usually removed by house bees, which is how these bees get the spores on them. They then spread the disease while feeding other larvae.

There is an antibiotic available to treat this disease, which is also used to treat another more deadly bacterial disease, American foulbrood (AFB). Beekeepers that routinely treat for AFB generally don't have EFB.

Because EFB is at its best when the colony is under stress, reducing stress is the best defense. Maintaining a strong colony, with lots of food, no other diseases, and a new healthy queen is paramount. Replacing old frames and keeping drifting to a minimum keeps the source down, too.

Avoiding Chalkbrood

> Strong colonies in the spring maintain even internal broodnest temperatures and can collect adequate food.
> Replace brood combs every three years or so to keep the number of spores to the very minimum.
> Remove and destroy entire brood combs that have lots of infected larva.
> Maintain bees that have hygienic behavior.
> As a matter of course, keep stress to a minimum with lots of available food, a healthy population, and a new queen every year.

European foulbrood-infected larva are first noted because they are tan to yellowish, turning dark brown and eventually to black. The larvae die before the cell is capped, and the remains are easily removed by the bees, thus somewhat reducing the incidence of further infestation.

American Foulbrood

American foulbrood (AFB) is caused by a bacteria, *Paenibacillus larvae*. This disease is by far the most destructive one honey bees get, and it requires the most drastic actions by beekeepers to avoid or eliminate it. AFB is spread by spores that are consumed by larvae during their first couple of days in the cell. After that, they become immune to the disease, but have already had ample opportunity to become infected.

You will probably never encounter this disease in your colonies. It is rare and well policed, but it is so very serious we need to spend some time with it, first, so that you can recognize it if it does show up, and second, so that you can efficiently deal with it so that it doesn't spread.

Spores enter colonies in a variety of ways:

> Workers rob a wild colony, another apiary, or another colony in your apiary, and bring home spores inadvertently through contact or in the honey they stole.

> Equipment previously contaminated is sold and reused.
> Frames are moved from a colony with AFB to one without it.
> Gloves and hive tools used when working a colony with AFB are not cleaned before use in another colony.

Ingested spores germinate in the gut, and the larva dies after the cell has been capped. The dead larva continues to provide food for the bacteria until it is mostly consumed. Once consumed, the larva is a jellylike mass that dries to form a hard scale containing millions of microscopic spores. These scales adhere to the cell wall tenaciously, and house bees most often cannot remove them. When trying to remove the scales, spores are spread throughout the colony; some reach the susceptible day-old larvae, and the cycle repeats.

Because of the seriousness of this disease, most state departments of agriculture have regulations on treatments and controls. Moreover, chemotherapy requires the use of regulated antibiotic drugs that, if misused, could enter stored honey and, inadvertently, be consumed by people.

This is serious stuff, and many beekeepers do not use the drugs, choosing rather to destroy contaminated equipment and bees. This treatment may seem drastic, but it is the most effective. When treating with drugs, if AFB is found in a colony, there are always spores available to start the disease again once applications of the drugs cease. You are, so to speak, locked into a routine of always having to apply drugs and never ridding the colony of the disease. Many beekeepers apply these drugs whether signs of the disease are present or not. These prophylactic treatments effectively stop the disease, but you must continue to apply them religiously. Properly applied, this is safe and effective in stopping the disease.

Drugs are applied to the broodnest area mixed with powdered sugar and are consumed by the nurse bees. Applications must be terminated *at least* 45 days before you install surplus honey supers to avoid any chance of contaminating honey to be harvested. As a result of this timing, fall applications are most common.

Drifting

When you have more than one colony, some bees from one colony will drift from home to the other colony. If you have options on where colonies can go, which way they face, and how close together they can be, you can probably reduce drifting between colonies. Drifting is a problem when more bees drift one way than back, weakening the donor colony. Diseases and pests can ride along, too.

To reduce drifting, face colonies that are close together in different directions, with entrances at 90° or 180° to each other. Using boxes painted with different colors gives the bees a clue of where to go, as does a landmark, such as a bush.

If you find that one colony is collecting a lot of drifters from a nearby colony, you can exchange the position of the two, helping to balance the population of both.

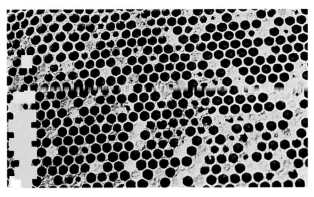

A spotty brood pattern is always a sign of trouble, and is a first, best sign of American foulbrood disease. This is your most obvious clue that a disease is present.

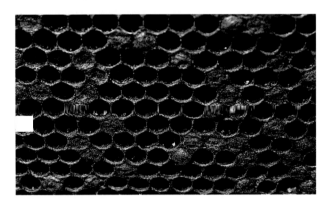

Look for scales, the dried remains of the dead larvae in the cells. They are hard and difficult to remove. Here, they are shiny black, but sometimes they are dull black. Usually the bees can't remove them, but they try, and in the process pick up spores spreading them around the colony.

There is an alternate choice. Careful examination and diagnosis can locate symptoms early so that brood frames, and entire supers if necessary, can be destroyed. Frames with diseased larvae can be removed and burned, but recall that the larvae don't die until the cells are capped, so detection isn't easy, especially when the infestation is small and early.

For the most part, you can assume that in even a mildly infested colony, most of the broodnest-area bees—those that feed and clean—have AFB spores associated with them. Further, these are some of the bees that take nectar from foragers, passing spores to them, also.

If you missed early detection, you can be fairly certain that most brood frames have several to many infected brood, that most of the nurse bees have been contaminated, and that some percent of the field force is contaminated also. This means when you destroy a colony's equipment, you must destroy the bees within. The value of early detection becomes more evident in this light. And the argument for prophylactic treatment with antibiotics becomes stronger, also.

Detecting early signs of AFB, or any brood disease for that matter, begins with examining the brood very carefully. Healthy brood is glistening white. Diseased brood—whether by AFB, EFB, chalk, or other diseases—are *not* white. The colors may range from translucent to tan to brown to nearly black. Your first clue, and the key to early detection, is the presence of larvae in their cells that *are not* pure, bright white.

Once an infected larva's cell is capped and the larva dies, the wax capping over the cell changes. Rather than remaining just slightly convex, it sinks in and will usually change color.

When applying antibiotic medications (there is currently only one, Terramycin, but others are slated for release) for American foulbrood, sprinkle it on the edges of the top bars in the broodnest. Avoid spilling it directly on the brood. The medication is mixed with powdered sugar so the house bees readily consume it, spreading it around the colony when they feed new larvae. The medicine keeps spores from germinating in the gut of susceptible larvae, protecting them from the disease, but it does not cure the disease itself.

Healthy larvae and pupa, such as those shown here, are pure, glistening white.

Avoiding AFB

AFB infects, on average, between 2 and 5 percent of all colonies in the United States every year. Considering the easy way it can be spread, this low rate can be attributed primarily to early detection by state departments of agriculture-sponsored inspection services, the use of prophylactic antibiotic drugs, and destruction of contaminated equipment by beekeepers.

Here are some tips for avoiding AFB:

> Never, ever buy used equipment—no matter how well you know the beekeeper, no matter how attractive the price, even if it's from your brother.
> Make sure your colonies are registered with your state inspection service (See Resources on page 160.) so they are inspected by a trained inspector who can help with identifying problems and suggest treatment options legal in your state.
> Seek out queen producers who sell queens that generate bees with strong hygienic behavior.
> Routinely inspect the broodnest (every ten days to two weeks, minimum) looking for spotty brood patterns, brood that's not glistening white, and sunken cappings.
> If available, contact your local inspector if you find suspect cells (see photos for examples).
> If AFB is confirmed, burn or treat forever and ever.

If you are going to burn infested equipment, dig a hole large enough to accommodate much of the equipment. Then, kill the bees by dumping a five-gallon (19 l) pail of soapy water directly into the colony from the top. Use one cup (60 ml) of dish soap in the pail of water. When the bees are dead, first put in frames and burn them, then add the boxes. When all the equipment has been consumed, cover the ashes completely. This is a horrendous loss of equipment, so avoiding AFB is certainly recommended. Before burning, check your local fire safety laws. A burning barrel also works for this.

Some house bees with hygienic behavior traits may begin to investigate this odd cell by opening it up at the center. Other bees with strong hygienic behavior will then remove the diseased larva before the bacteria have a chance to form spores.

This is where an effective pest-management program gets tricky. Because of regulatory zero tolerance for AFB, any infestation should be treated or destroyed. But a strong hygienic colony can tolerate a light infestation and clean it out. Nevertheless, knowing the telltale signs of an early infestation will make whatever control measures you choose easier and less destructive.

Tracheal Mites

Tracheal mites, *Acarapis woodi*, are microscopic in size and cannot be directly observed. These mites are considered to be universal, and unless you have known resistant strains of bees, you should assume bees you buy have them and you will have to manage them. Fertile female mites enter the main thoracic trachea (breathing tube) beneath the wings of young bees—queens, drones, and workers are all susceptible—lay eggs and produce young. Inside the tracheal tube (about the thickness of a human hair) the emerged mites pierce the tube and feed on the bee's blood, the hemolymph, that bleeds through the wound.

If infestations become excessive (and if left untreated, they probably will) bees die early, colony honey production suffers, and during the winter, most or all of the bees will die. Perhaps, in the spring, you will notice bees crawling near the entrance, but perhaps not.

Some strains of bees are tolerant or resistant to tracheal mite infestations, and you should definitely seek out these. Sadly, most queen producers do not produce resistant lines, and alternative means of control are required. Fortunately, these are easy to apply, inexpensive, and safe to use.

If you are not using bees known to be resistant to these mites, you can maintain a manageable level of these mites in your hives. You will probably never be free from them, but like weeds, a few are tolerable.

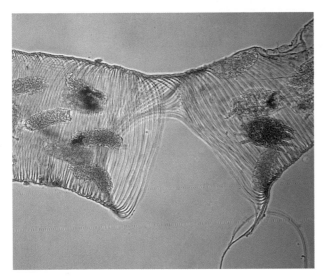

Tracheal mites live inside a honey bee's trachea. A trachea, or breathing tube, is roughly the diameter of a human hair.

Put the grease patty directly on the top bars, off center, and let the bees consume it. While doing so, they'll pick up minute amounts of the grease, foiling further mite infestations.

No matter where you live, an effective treatment called a grease patty is made as follows:

> Obtain a can of solid vegetable shortening—3 pounds (1.5 kg) is common.
> Get a pan large enough to easily hold twice this amount of shortening.
> Put the shortening in the pan, and slowly warm it on your stove.
> When the shortening reaches the translucent stage, and is not quite liquid, begin adding 10 pounds (4.5 kg) of regular sugar. (The formula is 3:1, sugar to shortening.)
> When all the sugar is added, add an additional ½ pound (.2 kg) of honey (from your bees or from a safe source). Stir in
> Turn off the heat, and add 1 ounce (28 g) of food-grade peppermint flavoring. Stir and mix well.
> Let cool.
> Using an ice cream scoop and waxed paper, scoop out about a hamburger-sized dose of the finished mix and put it on a sheet of the waxed paper about twice the dose's size. Add more paper and flatten. Freeze the patties until needed.
> Place one patty on your colony (leave the paper on the bottom) between the two boxes with the most bees in them, as early in the spring as you can, and replace it when most of it is gone. Continue adding these until your honey flow starts, then quit until the honey flow is over.
> Add a patty in the fall, and have one in place for overwintering.
> This recipe makes a lot of patties, so you may end up sharing with a friend.

Because sugar and shortening are essentially odorless to a honey bee, the honey and peppermint act as attractants. You will find some colonies eat these patties rapidly, whereas others are slow. If the bees are slow, add additional honey, sugar, and peppermint to a new mix to increase attractiveness. A rare few will never eat them. If that happens, leave them on anyway and hope for the best. It's all you can do for these bees.

Avoiding Tracheal Mites

> Purchase strains of bees advertised as tracheal-mite resistant. Some have been available for quite a while, whereas others are being selected. This is your best defense.
> Reduce drifting and robbing as much as possible.
> Keep grease patties in your colonies all fall and winter and into early spring. Remove in late spring until early fall.
> Requeen annually to keep strong colonies.
> Reduce other stresses.

Grease patties work well because the shortening, when on a young honey bee, confuses female mites looking for a new host, and the rate of infestation in young bees drops precipitously.

If you live in warmer areas, with winter temperatures that do not fall below freezing, you can use menthol crystals, available from bee supply catalogs. When placed in a colony, these crystals evaporate into fumes that are deadly to mites, but not to bees or people. Spring and fall applications may be needed, but never use them when honey supers are on the colony. Follow the instructions on the label. If the weather is warm enough, they are effective. If not, you've wasted time, money, and probably bees.

Occasionally, no matter what you do, a colony will succumb to these mites. This occurs most often during the winter and very early spring. You find in your colony lots of honey and few or no bees. Sometimes, you will find them in two, three, or more small clusters in different parts of the hive.

Remove the colony from outside after brushing away any remaining dead bees until you are ready to add new bees. But be certain the colony didn't perish from a disease (AFB) before using the equipment. There are no remaining mites in a colony such as this so recontamination on that count is not a problem.

Varroa Mites

Like tracheal mites, this pest is nearly universal, and like AFB, it is deadly without beekeeper intervention. Unlike tracheal mites, however, this pest is large enough to see, so detection isn't guesswork.

A fertile female varroa mite in your colony will seek out and find a cell with a larva that is just ready to be covered by the nurse bees. She scoots down to the bottom of the cell and hides there, in the remaining food and frass, until the cell is capped. Once protected, she climbs onto the pupa, inserts her piercing mouthparts into the pupa and begins to feed. Like many other mites, this blood meal is needed so she can begin laying eggs.

Like the tracheal mites, her firstborn is a male, who then mates with his sisters, who are produced next. One or two of these now-mated female mites, plus the mother, leave the cell when the bee emerges. Varroa females by far prefer drone pupa to workers because of the longer time they are in the cell, allowing them to produce more young, but workers and even queens are not immune.

Larvae that are attacked by varroa mites suffer all manner of problems. Varroa can transmit viruses that will stunt and eventually kill the newly emerged bee. Parasitized drones are shorter lived and less vigorous flyers than healthy ones. Workers, too, are shorter lived and have reduced feeding capacity, flying ability, and general health.

When young, newly mated female mites emerge with the drone or worker on which they had been feeding, they seek new cells to invade and repeat the process. The males do not leave

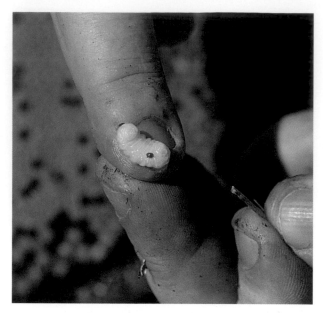

🐝 *If you remove a pupa from its cell that has a varroa mite already in it, this is what you will see. Note the size and color of the varroa mite. They are easy to identify. Generally, when you remove the pupa from the cell, the mite will scurry away, and they can move fast.*

but die in the cell. No matter what treatment choice you make, when the females are exposed, they are at the weak spot in their cycle. You must control varroa mites or they will eventually kill the colony. Fortunately, several treatments are available, and monitoring mite populations and treating is a routine part of varroa mite management.

Monitoring Mite Populations

Earlier it was recommended that you purchase a bottom board with a rear-access removable tray, and varroa mites are the reason. When female mites leave the cells, they walk on the comb and top bars and ride and feed on adult bees. They get from cell to cell, and even from colony to colony on drifting bees, in this way. Sometimes, however, they fall off the combs, or bees groom them off their bodies. And sometimes they just die and fall. Scientists have a pretty good handle on tolerable populations of mites in both small- and large-population colonies, and they know about how many will naturally fall in any given time period. You can use this information to your advantage.

The tray in your bottom board should have a mesh screen over it. Generally, the tray below is not in place, allowing the free flow of air for excellent ventilation.

In early to the middle of spring, take this tray and cover it with a piece of white paper, such as gift-wrapping paper. Lightly fasten this paper to the tray with a thumbtack or staple. Then liberally cover it with something sticky, such as spray-on cooking oil. Spray on the oil to the point of its running off the paper.

Mites that fall will go through the screen and land on this sticky surface and will not be able to walk away. Place the tray and sticky paper in the bottom board for three full days. Don't place the paper on a regular bottom board because the bees will have to walk in it and will clear debris at the same time, removing fallen mites.

After three days, remove the tray and count the mites (use a small magnifying glass, if needed), and divide by three. This gives an approximate average of mite fall per day. For a small colony in the spring (bees and brood in one or one and a partial box), the final number should be between 20 and 60 for a tolerable amount, and treatment can wait. If there are close to 60 mites, you'll need to watch so it doesn't balloon (and it may) before fall. An additional count may be wise in midsummer, just to be sure. If there are 60 or more, you'll need to apply a treatment.

This is a classic integrated pest management strategy, and if you are serious both about protecting your bees and being a conscientious beekeeper, this monitoring is important. If you treated the previous fall, a spring treatment isn't usually necessary, but the minute you take varroa mites for granted, they'll make sure your bees pay the price.

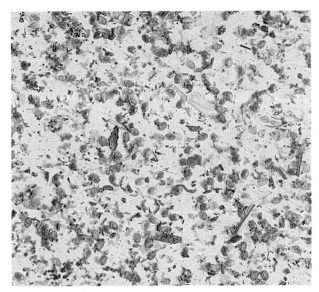

With a moderate to heavy varroa mite infestation in the fall, your sticky board will have 150 to 500 mites stuck to it after three days. Use a magnifying glass to help count the mites hidden in the other debris on the board. Once the count reaches 150 to 200 mites per day, you can stop counting. You've reached the treatment threshold and should not delay in applying medication or other controls.

Minimizing Varroa Populations

> Resistant bees:
Some strains of bees are tolerant of moderate levels of mites in a colony without suffering appreciably. Other lines apparently groom mites off their bodies (screened bottom boards are helpful here), whereas others restrict the reproduction capacity of the females when they enter the cells. Hygienic bees will remove some infested pupae.

> Screened bottom boards:
Whenever a live mite falls to a solid bottom board she will immediately reattach to another bee. If she falls through a screen to the ground below she is lost. Always use these screened bottom boards.

> Reduce drifting and robbing in your apiary, if possible.

> Remove drone brood:
Because female varroa mites greatly prefer drone brood for reproduction, you can provide frames of drone foundation (either wax or plastic) that the bees will make into drone comb and in which the queen will lay drone eggs. This separation reduces significantly the drone brood built on other frames. Placing one or two of these frames in a colony isolates the drone population and, thus, the reproductive mite population in your colony. To make this work, place a frame with drone comb in your colony. The queen will fill this with drones; mites will be very attracted to this frame and enter the cells. As soon as most cells are capped and the mites are trapped inside, remove the frame and place it in a freezer for a couple of days, killing the drones, and all the mites with them. This is incredibly effective in keeping the mite population to a tolerable level. Keep one in your colony from early spring until you shut down for the winter. You should have at least one in your colony all season if you are in the warmest areas of the world.

When you spray the bees with Sucrocide, the mites, but not the bees, are affected by the soapy water solution. The bees, when grooming themselves after being sprayed, knock off additional mites, which fall through the screened bottom board below. It is, to date, the safest treatment available and can be used even when you have honey supers on the hive (though you should not spray the honey directly).

Do a count again in late summer or early fall, but don't get caught treating too late. Here's what can happen if you do. Recall that varroa mites first seek drone cells. As summer wanes, queens produce fewer drones, and drone cells become fewer. Female varroa mites are forced to migrate to worker cells. Workers that have been attacked by varroa mites will seldom live through the winter. The result: a dead colony in the spring.

However, if the varroa population builds rapidly during the summer, these workers may perish even sooner. An out-of-control varroa population during the summer will invade essentially every worker cell, and if there are enough mites, two, three, or even six mites will invade a single worker cell, effectively killing the pupa before it emerges. When this happens, the colony will be dead before late fall. You can see the value of monitoring mite populations in the spring and summer to avoid this sudden and total collapse.

However, controlling varroa mites in a colony is fairly straightforward. And you have a lot of choices, ranging from no chemicals (and a bit of work) to pure pesticide control. The choice is yours, mostly, and they all fit nicely into a seasonal management scheme.

Remember to monitor mites on a frequent enough basis that you are comfortable with knowing the population. If the previously mentioned techniques fail to keep the mite population below the thresholds noted, you will need to treat the colony.

A colony with a very high mite population may get to the point of absconding. That is, the conditions in the colony

become so desperate that all the bees leave. They simply cannot cope and abandon the nest, usually in late summer. Where do these infested bees go? They migrate to the nearest colony. It may be a wild colony, one of your neighbor's, or even one of your own. This occurs almost always in mid- to late-summer. Then the population of the colony essentially crashes, and all you have are empty supers by late fall. Even a modest infestation in late summer can be fatal because affected workers don't live long enough to overwinter, and come spring the result is the same—a dead colony.

Treating for Varroa

The status of chemical treatments for varroa mites is dynamic, and old compounds continue to leave the scene with new products entering. There are some general recommendations that hold for most of them, but you need to keep up with current information in the journals, because some new miracle cure may show up eventually.

Soft Treatments

A variety of treatments using volatile essential oils are available or are being developed that offer effective control of varroa mites and, to some degree, tracheal mites. These treatments are applied to a colony and slowly vaporize, with the vapors killing adult mites outside cells but not harming bees. Mites in cells are not affected, and multiple or lengthy applications are needed to affect all mites in a colony. They offer, depending on time of year and level of infestation, around 70 percent or so control, which is good enough to keep things going.

These chemicals are not absorbed into the beeswax on frames, and leave behind no residue. However, you cannot treat your colony when honey supers are on because the honey will, to some extent, take on the flavor of the essential oil you are using. Read the label before using to make sure you are aware of all of the precautions.

Organic Acids

These chemical treatments are certainly evolving, but they have been used for several years in many countries. They are applied to colonies in absorbent pads and slowly volatilize, with the fumes killing mites but not bees. They can be tricky to handle but are effective. There are several—formic acid and oxalic are the most common—and several application techniques are popular, from gels to liquids. Like essential oils, these chemicals leave no residue in the wax and do not affect queen or drone reproductive capability as some of the harder chemicals are apt to. Organic acids will probably be registered for use in the United States, but many other countries are using them effectively. As always, read the label first.

Sucrose Octanoate

There is a nontoxic, benign sugar and soap solution that appears to be both effective and safe. It is called Sucrocide. It is so benign that it is registered for use in organic honey production. The principle is well tested. Soapy water solutions affect small organisms such as varroa mites (and aphids, plant mites, thrips, mealy bugs, and others) by clogging breathing tubes and causing dehydration. The trick is that the concentration has to be small enough to work in mites but not bees.

The sugar substance that's included induces the bees to groom themselves, removing additional mites.

The solution is sprayed, using a common garden sprayer, directly on the bees. Every frame has to be removed and all the bees sprayed, but once an assembly-line system is developed you can treat three brood supers in about 15 minutes. You need to do this at least two and usually three times to make sure you catch those mites in cells not yet emerged.

Hard Chemicals

There are currently two hard chemical treatments available to treat colonies with varroa mites. They are both very effective, easy to apply, and relatively inexpensive. There are, however, a multitude of precautions.

One of these chemicals is a synthetic pyrethroid, a familiar family of pesticides in the garden. The trade name is Apistan. It comes impregnated in a plastic strip and works similar to the flea collars you use on your dog or cat. The strip is placed in a colony, between frames. When the bees walk on it, they pick up trace amounts of the active ingredient—fluvalinate—which then kills any mites they contact.

The second hard chemical available is an organophosphate, with the trade name Checkmite+. It, too, is impregnated in a plastic strip and works essentially the same way as Apistan. The active ingredient is coumophos, often used as an external flea and tick treatment for livestock.

Both of these chemicals, as strips, are placed in a colony long before or immediately after honey supers are on your colony. Strips are applied, one for every five (deep) frames of bees and brood, in the broodnest supers. Check the product's box for how to figure application rates.

The strips are left in the colony long enough that all exposed mites and any that might emerge are exposed to the poison. Then the strips must be removed.

Herein lies the precautions when using these chemicals. They are effective and will save your colonies from varroa mites when used correctly. Their legacy, however, is less positive. Both of these chemicals are readily absorbed by beeswax and eventually the buildup in the beeswax begins to affect the bees. The so-much-and-no-more rule is broken, and the balance of enough to kill mites and not enough to harm bees goes askew.

A Checkmite+ strip hangs down between frames in the broodnest so bees can walk on it and pick up minute amounts of the active ingredient. Don't put it on the bottom board, on the tops of the frames, or anywhere but between frames. The strip is held in place by the fold-out tab on top.

🐝 *Adult wax moths are about the size of an adult worker honey bee.*

🐝 *As the larvae eat their way through the comb, they leave behind webbing, which hinders the bees from catching and removing them and from being able to use the comb, or to even clean it.*

🐝 *Wax moth larvae, like the one shown here, do great damage in a beehive. The larva is an off-white, soft-bodied caterpillar. This one is about half grown.*

🐝 *Cocoons of wax moth pupae are very tough, and bees usually can't remove them.*

Once and certainly a twice-per-year application for more than two years accumulates enough of either of these materials to adversely affect the queen's egg-laying ability and the fertility and health of the drones your colony is producing. The price of this excellent varroa-mite control is comb replacement every two or three years. If you choose this route, you must follow label instructions precisely, replace the comb on a regular basis, monitor populations, and use alternative controls as often as possible.

Wax Moth

Wax moths, *Galleria mellonella*, (also known as greater wax moth) can be a real nuisance, but they can be taken care of pretty easily.

Sometime during your bees' first summer, a mated female wax moth will in all likelihood find your colony. She'll get inside by sneaking past the guard bees, usually at night. Once inside, she will lay eggs somewhere in one of the boxes with brood. The eggs hatch, and the moth caterpillars begin to feed on beeswax, pollen, honey, and even larvae and pupae, that is, unless house bees catch and remove them. If the colony is strong and healthy, the internal police force is very protective against these invaders. But small colonies, or those being stressed by other troubles, aren't as diligent, and wax-moth larvae can make some inroads.

If they get a foothold, the moth larvae tunnel through your brood comb, leaving webbing and frass everywhere in their paths. Unimpeded, wax moth larvae can, with favorable temperatures, completely consume the comb in a super in ten days to two weeks. They eventually pupate, spinning tough cocoons fastened to the sides of supers, top bars, or the inner cover. They literally chew a groove into the wood so they fit. The cocoons are

Avoiding Wax Moths

> Don't pile too many empty supers on top of a colony; only use enough to allow the bees to keep moth larvae in check. You may find one or two larva in a small colony, but they should be hard to find in a large colony.
> Make sure colonies are healthy. A stressed colony will have enough going on without having to deal with moth larvae.
> Do not store empty honey or brood supers in moth-friendly environments—warm, dark areas with plenty of food, such as your basement or garage.
> Wax moth larvae do not thrive when exposed to light and fresh air. A stack of supers with a secure top piled in your basement is the perfect place for these pests' populations to explode.
> If feasible, stack unused supers on their sides with a few inches between them, and supers on top oriented at 90°. This placement allows light and fresh air into the super, greatly reducing larvae activity. You can also build a rack that exposes the supers while protecting them from rain. Expose, expose, and expose your supers to keep moth populations manageable.
> You can keep a super or two on strong colonies after harvest and let the bees handle the worms, if you live in moderate to cool regions.
> For a few supers, place the whole super in a freezer, set at 0° F (−18° C) for 48 hours. This kills eggs, larvae, pupae, and adults. Once the outside temperature goes below 40° F (5° C), the temperature essentially halts all moth activity (but does not eliminate them), and your supers are safe for the winter, no matter where or how you store them, as long as it stays that cold.
> If outside storage or freezing isn't possible, as a last resort, you can put a moth fumigant on a stack of supers. The only chemical approved for this is a formulation of paradichlorabenzene available from bee supply companies. This is the same chemical in crystal form used to protect clothes from destructive moths, but without other fragrances and additives. Do not use any other formulation of this chemical—it is a violation of the label and may damage your combs. And then use it as a last resort and only sparingly. The wax will absorb fumes. Before using next season, set frames outside one full week to air out and let as much of the material evaporate as possible.

so tough that bees can't remove them. Adults emerge, leave the colony, mate outside, and then the females find more colonies to infest. They'll do the same to supers you store in the basement or garage, unless you take some precautions.

You can be pretty sure there are always some wax moth eggs in your colony. Adult female moths routinely enter and lay eggs, but the tiny larvae are just as routinely removed by the bees. Thus, when finished using a super at the end of a season, those eggs are still there, but the moth police have gone.

Wax moth adults are around pretty much all year in tropical and semitropical areas, and most of the year where there are mild winters. In cold climates, wax moths are a problem only from midsummer until hard frost—outside, that is. Inside the hive is a different story.

Chilled Brood

An environmental problem called *chilled brood*, occasionally occurs when a rapidly expanding, healthy colony produces lots of brood in the spring. If the weather turns suddenly very cold, the bees will cluster in the very center of the broodnest to keep each other warm. When this happens, uncapped brood on the very edge of the broodnest will not be covered and will drop

Here is an easy-to-build storage rack for your extra supers when they are not on a colony. This arrangement allows light and fresh air into the supers, discouraging wax moths from infesting the combs. It also protects the supers from the rain and other elements.

🐝 *Dead larvae or pupae found on the ground outside the hive after a cold snap usually indicates chilled brood. These disappear fast though, because ants and birds will eagerly remove this free meal.*

Shown is a small hive beetle adult. Screened bottom boards help remove the larvae of this destructive pest, in-hive traps can reduce adult populations, and sprays applied to the ground outside the hive can help keep pupating larvae from returning.

below the 95° F (35° C) temperature the brood need. If this lasts overnight, some or perhaps all of the exposed brood will die of exposure.

The next day, you'll see blackened brood of all ages at the edge of the broodnest, whereas those in the center remain healthy and glistening white. Within a day, the bees will remove all of this dead brood, and you'll find them on the ground just outside the front door. All ages of larvae will be there, and this is your clue to what happened. You can't prevent this, but if you are aware of the situation you won't spend time treating a problem that doesn't exist.

Small Hive Beetle

A relatively new pest in the United States is the small hive bee- tle. An import from South Africa, it has no competition in the United States and caused significant problems when it was first introduced. It has become established in the very southern parts

of the country, and has moved north along the East Coast in the sandy, soft-soil areas. It has spread to other parts of the country, but because it is a tropical insect, it has not become a problem anywhere else.

Adult female beetles enter a hive and lay eggs. The beetle larvae tunnel through comb, eating pollen, larvae, and honey. As they progress, their excrement causes stored honey to ferment and bubble out of the hive in an awful mess. When the larvae are mature, they leave the hive, burrow into the soil, and pupate. Adults emerge from the soil and infest other colonies, or rein- fest the original colony. Strong colonies can usually hold their own, but weak colonies and honey supers stored for extraction can be destroyed in a matter of days.

Removing an infested super from a colony prevents further damage. In the most tropical parts of the United States, available controls are less than adequate, and inspection and ground treatments are all that's available. Read the journals on newer management strategies for small hive beetle, treat pupating sites adequately, keep colonies in the sun and away from forested areas, and use the available in-colony traps. Effective lures and traps are being developed, and as one researcher said, these beetles will very soon be a nuisance rather than a pest.

Animal Pests

If your colony is raised off the ground a couple of feet, skunks and opossums are seldom a problem, but continue to check for them. Skunk visits are noted by torn-up sod or mulch directly in front of the hive, and muddy paw prints or scratches on the landing board. A skunk will, if it is reachable, scratch at the hive entrance at night. Guards who investigate are grabbed and eaten. Other guards will fly and sting the intruder, but skunks are nearly immune to stings on their paws, face, and even inside the mouth. This can last for some time during a night, and for many nights in a row. A mother skunk will bring her kits and show them how to harvest this sweet, high-protein snack. A colony that's attacked will become very defensive because of the constant disruption and continuous exposure to alarm pheromone, especially the day following the attack. Opossums are generally opportunists and grab what they can without the scratching.

Raccoons may also investigate your colony. They are usually attracted to a hive because of wax or propolis carelessly dis- carded in the vicinity of the hive. Here's a word to the wise: Raccoons don't attack the front door. They will, if determined, remove the cover and inner cover (still loose from your recent inspection) and pull out a frame. They'll drop it to the ground, and then drag it away several feet. Guard bees or any bees on the frame fly or crawl back to the hive, leaving the raccoon to feast on honey and brood in relative peace. Placing a brick on top of your hive satisfactorily prevents this.

Summertime Chores

By early summer, your brand new package colony has most likely grown past the rigors of becoming established. The stresses of temperature fluctuations are reduced, the population is building, the queen is producing, nectar and pollen are being collected, frames are mostly built up with beeswax cells, and honey is being stored. This is the norm.

But if you have a dozen colonies, one or two may not be thriving and need some additional attention. Having at least two colonies is recommended, so you'll have a basis for comparison. How can you tell what's happening, no matter where you are? The bees don't send out press releases, but their activities are indicative of their situation. Mostly bees react to their environment rather than plan ahead. You should, however, be a step ahead of them so you can anticipate their needs. Temperature extremes, rainfall, and even weed growth are limiting factors for your garden plants, and for the plants your bees visit. But even with that experience, an occasional chat with nearby beekeepers can be enlightening, and belonging to a local club becomes even more valuable. An experienced local beekeeper can, in a few moments, share the typical season's progression—that bit of wisdom is worth its weight in gold.

At the same time, keep your record book up to date. Your bees are exploring their environment and finding, or not finding, sources of food. Until you know as much as your bees do, you're still learning.

Your advantage, of course, is that you can keep records. A colony of bees has to reinvent the wheel every season. The few bees that overwinter were born in the fall and have no institutional memory, as it were, so the colony as a unit must learn it all again every year. You, on the other hand, can quickly review last year's records and recall flowering times for fruiting plants. Being aware of how the season progresses, whether short and ending early, or the longer season of moderate climates allows you to anticipate swarming, honey flows, and preparation for the slow or winter season. Your

package colony will be playing catch-up most of its first season, but it will be reacting to the local environment in much the same way as more established colonies.

Your package colony, in eight to ten weeks after installation, will probably have brood in all three of their brood chambers, but that will range from a box and a half to three boxes full. The difference will be dependent on how favorable the weather has been, the type of bees you have, and any obstacles that have arisen during that time.

By midsummer, honey will probably be in at least one and perhaps part of a second honey super. Looking ahead, you must prepare for the slow season—cold weather in the cold regions, and the rainy or cool season in the warmer areas—and make sure your colony has enough honey stored to last until spring returns.

In the coldest regions, where snow is common and winter lasts six months or more, a typical colony will need about 60 pounds (27.2 kg) of honey to survive. An eight-frame medium super will hold, if completely full, roughly 30 to 40 pounds (13.6 to 18.1 kg) of honey. In the warmer areas, 40 pounds (18.1 kg) is about all you'll need, and in the warmest areas, where forage is available year round—southern Florida, for instance—additional food is not required.

By midsummer, honey becomes an issue for you and your colony. How much do you want? And how do you manage that? Certainly nature may make that decision for you, but during an average season, your package colony may make 40 pounds (18.1 kg) of surplus honey. It can make more in a good year, but maybe none during a poor year.

You must provide room for honey production—room to dehydrate nectar—using an additional super or two. Deciding how many supers to provide is definitely an art. If you provide too much room, the super will remain unused, which isn't too great a problem the first year. Your bees will store some honey in the broodnest boxes. You want them to store enough but not so much they fill space the queen needs to continue producing the

Surplus

Surplus is the term beekeepers use for the honey they harvest. It is anything more than the 40 to 60 pounds (18.1 to 27.2 kg) the bees need to overwinter. Your package colony—playing catch-up because of having no stored honey, having to build the entire wax comb, and raising young—starts the season essentially at a deficit.

To store surplus honey, your colony needs to first produce the equivalent of about 60 pounds (27.2 kg) of honey plus continue to feed a growing population of young bees. Your package bees have to really hustle just to stay even. And, if the season's off, they may never get even, and you'll need to feed them most of the summer. Next season, they'll have a reserve of food to feed a growing population and won't have to produce nearly as much wax.

next generation of workers. In fact you must prevent this by moving frames full of capped honey out of the broodnest and up into the honey storage supers.

To get the necessary amount stored, you want that 40 pounds (18.1 kg) in one super and an additional 20 pounds (9.1 kg) stored around the edges of the broodnest. That's about eight full frames of honey out of the 24 you have in the three brood boxes. If there is more, you should move some out. If there is less, you either move some in or reduce available storage above the queen excluder.

If you are having a bumper crop, you need to continue to add boxes, two or three, as the first boxes added are filled. Here's a trick if your colony fills those bottom three boxes with almost all brood and stored pollen and keeps putting honey above their excluder. Take a nearly full, partially capped honey super and place it on the very bottom of the colony. This is an abnormal situation for the bees, and they will actually (almost always) move it up into the broodnest. This solves two problems: not enough honey in the broodnest for overwintering and not too much above. During the summer, don't get too carried away adjusting honey storage. Usually the bees figure it all out without too much interference, but keep an eye on it.

Late Summer Harvest

Your first-year colony probably won't have a harvestable amount of honey before late summer, if then. The conventional procedures for harvesting are outlined later and are simple enough to carry out alone, and even easier with help.

But now, preparations for winter begin in earnest, and they need attention. As day length shortens and weather cools, the queen's productivity slows and there will be less brood. Your drone frame may be mostly empty, but keep it in the hive until you harvest, replacing it at season's end with a full frame of honey for overwinter stores. Check the brood nest very carefully for signs of disease, because this is when problems you may have missed earlier begin to emerge in a first-year colony.

There will probably be incoming nectar for a bit, but examine closely the amount of stored honey, making sure there's at least the 60 pounds (27.2 kg) needed for the slow season. Be mindful that a first-year colony may need additional feeding to supplement that ration and be prepared with feeders and sugar.

Removing Honey

If at the end of the season, before fall sets in and before you apply any necessary medications, your colony has *surplus* honey, it's time to harvest. Countless beekeeping books explain nearly countless ways to do this, but for a hobby beekeeper with only a very few hives, neighbors, and not much time, there's only one good way to do this.

Fume boards are easy to make. This one is made from scrap pieces of 1" × 2" (2.5 × 5.1 cm) lumber and plywood. The inside is lined with something as simple as thick, absorbent cardboard. You can purchase preassembled models that come with cloth liners.

Using a Fume Board

To remove a box with frames full of capped honey and leave the bees unagitated in the colony below, you must use a fume board. This simple device looks much like a telescoping outer cover, except it is the exact outside dimensions of a super, not larger, so it sits on top of the super below it. You can make or buy one. The principle behind this technique is fundamental. On the inside of the fume board is an absorbent pad. It can be flannel, cardboard, or wood. You apply the correct amount of a chemical repellent to the pad, put the fume board on top of the super full of honey (cover and inner cover removed) you want to remove, and wait 10 to 15minutes for the bees to move down into the colony away from the fumes and out of the honey super you want to remove.

The chemical product to use has the brand name of Bee-Quick®, and is a pleasant-smelling (to humans), nontoxic concoction with a vanilla fragrance. If it isn't already permitted to be used in organic honey production, it probably soon will be. It's that safe and easy to use.

Fume boards are usually dark colored or black (furnished that way or painted by you), and on a sunny day they will warm the repellent, vaporizing it, and flushing the bees out of the super. In a few moments, the bees will be gone, and you can remove the bee-free super and place it in a bee-free location.

The reason this fume board works so well is that your neighbors will not even suspect you are harvesting. If correctly managed, no bees fly, no bees become defensive, and no bees even seem concerned. However, too much of a good thing can and will cause a problem. Apply just the small amount of repellent recommended by the label. If you overapply, you will chase bees out of one, two, maybe three supers. And they'll run right outside. Suddenly you will have a flood of bees pouring out the door, stumbling over each other to escape. This is not good. Err on the side of not quite enough at first. You can add more if needed, but you can't push bees back inside the hive.

You can't just leave surplus honey in the colony over winter. It will most likely crystallize in the cells. Cool temperatures hasten crystallization, and those frames will sit on a colony, uncovered by the cluster, unprotected from wax moth, and turn hard as a rock.

When this happens, you can't get the honey out, no matter what you do, and neither can the bees. Come spring it's still there, and neither you nor the bees can use the frame.

Your bees need some honey, certainly. Bees will consume 40 pounds (18.1 kg) if it's warm all winter, maybe as much as 60 pounds (27.2 kg). A medium frame full of honey has about 5 pounds (2.3 kg) of available food. To supply 60 pounds (27.2 kg), use 12 full frames of honey. Often, but not often enough to gamble on, a first-year colony will have six frames of honey somewhere in the three brood boxes. It may be 12 half-frames or some other combination. Examine them so you know. That's 30 pounds (13.6 kg) or so. There may be more, so look carefully.

This frame has brood in the center and honey on the edges and sides and top. The honey stored here will feed brood during the winter.

A full super of honey frames added to that will give your bees 70 pounds (31.8 kg). Or, they may have more below and need less above. This arithmetic is necessary.

Don't guess about the amount of honey to leave the bees. Too little and they will starve in late winter, when the need for lots of food to feed lots of new brood escalates. In fact, late winter is the time of year that most colonies perish—either because they simply run out of food, or workers raised in the late summer were damaged by tracheal or varroa mites.

Several things influence the actual date you will remove honey from your colony in preparation for the coming fall and winter seasons. Foremost of these is the need to treat for varroa mites. If you are using any of the hard chemicals or the essential-oil potions, you must not expose honey that's intended for human consumption to these chemicals. So, treating those bees raised in the fall for varroa mite infestations influences when to harvest honey.

Another consideration for harvest time is your time. If you are removing honey and giving it away, your time commitment will be minimal—perhaps an hour on one weekend day. However, if you'll need to prepare a space—the garage or the basement—harvest and

Beware of Foul Fumigants

There are other chemical repellents that work, but think twice before using them. They are toxic, flammable, and the foulest smelling concoctions ever created. They are effective (Bee-Go® and Honey Robber® are their names), efficient, and will not taint honey or equipment. Spill them on your bee suit, in your car, or in the house, however, and you will be sorry forever or until you move or destroy the fouled items.

move those frames there, extract, remove the frames, and clean up. It can take the better part of a couple days with one or two colonies. Plan accordingly, and factor in when treatments will need to begin. Your record book will help here, especially next year.

Removing Frames or a Super

The best advice for learning to harvest honey is to help someone else do it before you have to do yours. But that may not be possible, so here are some hints and tips for making this task as easy as possible.

If time permits, the day before you harvest, quickly examine just the honey supers you will be working in the next day. Smoke the bees a tiny bit so they leave the super and go below. Then quickly loosen every frame with your hive tool, breaking all bridge and brace comb and other anomalies. Loosen, too, the whole super so you can move it easily the next day. Overnight, the bees will clean up any honey from the broken comb and you won't have leaky, sticky frames to handle and store.

This quick exam also shows just what you'll be harvesting, so you have the right equipment ready. If the super you have has surplus honey in it (additional to the amount needed by the bees), whether completely full or with only a few frames, here's how to remove them quickly, easily, and safely.

> Assemble everything you'll need, check the weather and your neighbor's activity, and if all's clear, get everything to the colony.
> If you'll be pulling only a few frames, put your container next to your colony, with the lid off and nearby.
> If removing a whole super, lay down the extra cover or plastic and have the top ready, on the hive stand or in the cart.
> Prepare the fume board.
> Squirt or spray just barely enough of the repellent on the absorbent material inside.
> Err on the side of too little rather than too much—you can add more, if necessary.
> Don't smoke the front door.

 Get all your equipment together before you begin. Use a plastic, tote-type container to transport individual frames of honey from the beehive to wherever it is that they will be processed or packaged. Remember, honey is a food and should be treated with respect. Keep everything it touches clean and tidy. Your family may be eating it tomorrow.

> Do remove the cover, smoke, wait, remove the inner cover, and smoke again. You want the bees to begin moving down to the super below.
> Waft two or three puffs, and place the fume board on top.
> Wait
> Wait a little longer.
> Check beneath the fume board to see if there are bees there.
> If not, gently lift the super and look at the bottom. If you see bees, replace the fume board and wait a bit more.
> If after 10 or 15 minutes there are still bees on the bottom, add a tiny bit more repellent to the fume board.
> Wait
> When the bees have cleared the super, remove the frames to be harvested or the whole super. Leave frames with unripe honey. (Ripe honey is covered with wax cappings.)

Harvest Equipment You'll Need

> Fume board and repellent. Pick a warm, sunny day if at all possible. Make sure your repellent container is open and unsealed before you need it.
> Smoker, hive tool, protective gear, extra frames, and varroa treatments, if using.
> Bee-tight container to put frames in—a plastic container large enough to hold as many frames as you'll need (up to eight) with a tight, sealable lid.
> If removing an entire box, a bee-tight bottom and cover—two covers work if you have them, but two pieces of plywood or heavy-duty plastic sheeting will also work.
> If the supers are too heavy or too far to carry, you'll need a cart.

> Put full frames in the container and cover between adding another frame. Don't start a robbing melee!

> Replace frames removed with empty frames. Don't leave an empty spot, even for a few days. Put frames with foundation only on the edges, moving frames with comb to the middle.

> If removing a whole super, place it on the board or plastic and cover it immediately.

> Apply varroa treatment, if planned.

> Close up the colony.

> Move the honey and gear inside, and pat yourself on the back.

A good place to store a few frames is in the freezer if you aren't going to deal with them immediately. Otherwise, get them to the place they will be handled as soon as possible.

If you are disposing of surplus honey (uncommon, but don't count it out) double-bag it, so it won't be robbed and put a lid on the trashcan. All manner of animals will be interested if you aren't. A better solution is to simply give it to another beekeeper to extract. He or she will either keep or share the honey. Either way, the task is resolved.

Harvesting Honey

Bees store honey. During a good year, they'll fill every space you provide. Honey is the primary reason many people have bees. Here's where you have to take an honest look at this part of having bees. It reflects back to the zucchini complex mentioned earlier, but it applies to any garden harvest. How much honey can your family eat or use?

Honey yields vary every year. The amount of rain, varmints ransacking the place, temperatures that are too cold or too hot, too little attention—all are factors that affect how much or how little honey you'll get. But most years, your colony will produce between 40 and 60 pounds (18.1 to 27.2 kg) of harvestable honey. Some years, with adequate attention, 100 pounds (45.4 kg) isn't out of the question.

How much honey is that? A typical 5-gallon (19 l) pail holds 60 pounds (27.2 kg) of honey. That's not a lot to distribute. If you're raising comb honey, that's 10 or 12 round combs frames, or 120 or 130 of the small rectangular containers.

If you are realistic about how much to expect before you begin, you can plan the type of honey to produce, how to best manage for production, and how to process it and handle it once harvested. This goes a long way in avoiding the zucchini complex.

Extracted Honey

This is the most common form of honey, the one you're already familiar with—the kind sold in a jar.

To produce liquid honey, you give the bees frames with plastic foundation. They build beeswax comb on the foundation, fill the cells with honey, and cover them when the honey is less than 18 percent water (called ripe honey).

You'll find frames that have what are called "wet" cappings, as shown at top, and "dry" cappings, at bottom. When bees place the wax covering over the cell filled with ripe honey, they either place the wax capping directly on the honey, giving the cap a wet appearance, or they leave a tiny airspace between the wax and the surface of the honey, giving the cap a dry appearance. Comb honey producers prefer the dry look, but neither wet nor dry caps have any effect on the quality or flavor of the honey.

Then, you remove the frames, and remove the wax cappings, leaving almost the entire honey-filled comb intact on the frame. You'll save the beeswax and honey you remove for later use.

The honey-filled frames are then put into a machine, called an extractor, which is powered by an electric motor or muscle power. The extractor spins the frames at a fairly high speed, so the liquid honey runs out of the cells and is collected in the bottom of the extractor. It works just like a salad spinner, except that you drain the honey into jars or pails. The frames are given back to the bees for a time, so that they can clean up the sticky bits left in the box.

Extraction can be done whenever there is ripe honey to be removed. Most people who extract honey tend to bunch their activity so that the setup, extraction, and cleanup time is efficient. Also, larger extractors need a minimum number of frames to be full and to run.

One way to handle this task is to take your supers to another beekeeper who extracts honey at the same time you do and combine your frames to fill the extractor. All manner of negotiations occur for this service, because that's what it is—a service.

🐝 *Uncapping knives range from heat-controlled units to heated units without controls to unheated uncapping knives (uncapping knives have offset handles so you don't bump your knuckles on the frame as you remove the wax) to simple serrated kitchen knives, which work well for processing 20 to 50 frames.*

🐝 *A practical uncapping tub has a large container on top to catch cappings and hold uncapped frames until ready for the extractor, a metal grid to support the cappings, and a holding tank below into which honey can drain. A valve drains honey from the bottom tank. A removable mesh liner filters the honey before it drains below so you don't have to filter it again.*

🐝 *A hand-powered extractor holds four frames at a time.*

🐝 *An extractor with a small motor makes the task much easier.*

🐝 *Cut the cappings off using a knife, and let the capping fall into the tub.*

An uncapped frame, with honey ready to be removed.

Drain the honey from the extractor and strain it to remove bits of wax, propolis, the errant honey bee, and the like. Many types of strainers are available. Most fit right on 5-gallon (19 l) pails. Honey can be conveniently stored in reusable, clean 5-gallon (19 l) pails with lids. You can't have enough 5-gallon (19 l) pails if you are in the honey business.

Assisting with the work is a good way to learn the task without buying the equipment first, and two always work faster than one.

Leaving some amount of honey may be part of the deal, along with the cappings wax, or there may be a simple per-pound charge for the task. Often you can purchase beeswax that has been cleaned and melted, avoiding that task also. Of course it's not your wax, but beeswax is beeswax, after all. This isn't a free service, so don't expect that. But with only two or three colonies, justifying the cost of all the equipment needed to extract and process this small amount of honey and wax can be difficult.

Even if you can't locate someone to do this, visit a beekeeper or two who already have the equipment. The whole process is complicated enough that a bit of study is warranted before you begin. Take notes and pictures.

Once you have a feel for the workflow design and the equipment you'll need, where will you set up all this at your house? A basement is often used, because it's inside, away from the elements, out of the way of others, and there's running water to clean up with. A walk-in basement can work well because you're not carrying supers down and then back up stairs.

Kitchens are, or should be, out of the question. Wax, honey, propolis, and bees in the house seldom make good impressions.

The garage, as long as it's relatively bee-proof, is often used, and you can clean it later.

The first equipment you need to consider is a way to protect the floor, or better, a floor with a drain, which allows you to thoroughly wash the floor after extraction.

Supers with frames full of honey are brought in and set on some sort of catch tray—an overturned telescoping cover works—to keep drips from running out.

If you're doing the extraction work alone, think through the process. Cappings are removed using a variety of tools but most commonly a type of large knife. For only a few frames (two or three supers), a serrated bread knife is usually sufficient. For more frames, a specially designed knife should be used, and for numerous frames (10 or more supers), a heated knife is best.

The cappings are cut or scraped off into a tub or specially designed tank. Then the frames are placed in the extractor, spun, removed, and put back in the supers. Each step needs a plan. Where do the frames go *after* they are uncapped but before going into the extractor? They'll fit into one of those tubs, if you have one. What if two people are working together? Is there room? Can the equipment handle double speed?

When planning your workflow, remember the very true, very sad observation made years ago by Ohio's Extension Specialist for Apiculture, Dr. James E. Tew: "Most people get into beekeeping because of their curiosity about bees, but they leave beekeeping because of the nightmare of harvesting honey." *Nightmare* may be a bit harsh, but without planning, it can be a headache.

Dealing with Cappings Wax

If you extracted your honey crop, you'll end up with two products: the honey, of course, and the beeswax collected when you removed the honey comb cappings prior to putting the frames in the extractor. You'll now see the wisdom of lining your uncapping tub with a filtering cloth available just for this purpose. By gathering together the edges of the cloth you can lift the whole mixture out of the container in a perfect filter sock.

Suspend this filter sock over one of your 5-gallon (19 l) pails. The honey will drain out and can be added to your harvest, leaving behind only the sticky, yet beautiful, beeswax to finish later. You can certainly leave the wax and honey to drain right into the cappings tub, unless you need it right away. In any event, don't leave it there for more than a few days. While it is draining, keep it in a warm place and keep it covered. More important, keep it in a bee-free environment, such as a warm garage or basement. Draining the mixture to separate the honey will take a couple of days, so be patient.

The next step is to melt the beeswax for further use. This simple process is often referred to as *rendering* the wax. Before beginning, you will need to organize your utensils. First, a large container, a 2- or 3-gallon (7.6 or 11.4 l) cooking pot not used for anything else (not made of copper or iron because these will stain the wax), a heat source (an inexpensive gas or electric hot plate is ideal for this), a dipping pan or metal ladle, some filter material, and several smaller containers for your final product—beautiful beeswax.

As quickly after harvesting as possible, melt the cappings so they are ready to use later and so that wax moth larvae don't get into the wax. If you are unable to take on the melting process immediately, you can store the mesh bag containing the cappings in a plastic bag in your freezer.

When you decide to melt down the cappings, find a place that will be immune to a few wax spills. There will be some drips and drops no matter how careful you are, so be prepared. Cover your work surface with a few layers of newspaper or other covering. After the first few times of doing this, you'll have a better feel for how messy this task is, and you can prepare accordingly.

All you need to melt your cappings. A heat source—here a small, portable propane hotplate—a large enamel kettle, and a ladle.

Properties of Beeswax

> Beeswax melts at about 145° F (62.8° C). This temperature will vary a bit depending on air temperature and amount of debris in the wax.

> Density is about 0.96, whereas water is 1.0, so wax will float on water.

> Cappings wax, when cool, will be a soft lemon yellow color. Wax from old frames and bits of burr comb will be darker and will contain melted-in materials such as propolis. Do not mix cappings wax and old wax.

Cleaning Up Wax Spills

> Wipe up fresh spills with paper towels.

> Scrape up cooled drips and dribbles with a sharp-edged tool, such as a single-edged razor blade.

> To remove small spills and the thin film that remains after wiping and scraping, use a petroleum solvent specially made for wax removal available at most stores that sell candles. A final rinse with hot, soapy water will finish the job.

An old enamel coffee pot is perfect to catch wax in if you are going to pour it again. This one has a paper towel filter. Note the color of the wax. This soft lemon yellow is the highest quality, most sought-after color for beeswax. This is what melted cappings will look like.

Paper milk cartons work well for holding beeswax until it cools and is ready to be made into candles, creams, or soap.

CHAPTER 4 ✦About Beeswax✦

A huge selection of molds is available for making candles for every taste and occasion.

Making Candles

Candles made from pure beeswax are unsurpassed in fragrance; they provide long, clean, smoke-free burning; and simple beauty. Numerous books and reference materials on making all types of candles are available. If this is a part of beekeeping that you enjoy, you'll want to explore more than the simple candles shown here. But these fundamentals for making candles provide a great place to start.

Before you begin, you'll need some supplies, notably the molds in which you will make your candles and the wick that is the heart of every candle. Many companies sell all you'll need, and you can find some in the beekeeping journals, on the Internet, and in candle books.

The wicks you'll need come in various sizes, shapes, and materials. The primary consideration is the diameter of the wick. If it is too large, the candle will burn with too large a flame and smoke; if it's too small, it won't burn at all. And to confuse matters, different catalogs use different measuring systems. Most will give the size (diameter) of candle a particular wick works best with, but some don't. Get several catalogs for comparison before you begin.

Mold designs are infinite, so finding what you want—from elegant tapered dinner candles to a candle shaped like an ear of corn—is fairly simple. The most popular and most usable molds are made of polyurethane. These are split up one side so the finished candle is easy to remove. They are held snugly together during pouring with rubber bands so that wax doesn't leak out, and any seam that remains can be easily removed. Some shallow molds don't need the slit—you can simply push out the finished candle. These easy-to-use and fairly long-lived molds are great for beginners.

No matter what you want to make, you need clean wax. If yours hasn't been through the final cleaning, now's the time to finish this simple task.

Prepare your surface, covering it with newspaper to protect it from spills. Make sure your electric cords are out of the way, and the area is free from extraneous items you might bump with a ladleful of hot wax. Arrange equipment for easy pouring and moving. Don't leave hot wax unattended. Make sure you have a fire extinguisher handy.

Get out your hot plate. An old enameled or aluminum coffee pot is ideal for this process because it pours the melted wax easily, but any pot will do. Set a shallow pan partially filled with water on the hot plate. Into the pouring pot, put a piece of the wax you want to clean.

◀▮▶

Choose the correct size wick. This may take some experimentation, but it's all part of the learning process.

To clean your wax, warm it in the first pot until it is about 180° F (80° C). While it is warming, secure your final filter to the second pot. This pot *is not* yet on the hot plate. A double layer of paper towels is a good filter for small batches of wax.

Pour the melted wax from the first container into the second. You'll now have clean wax for candle making. If it will be a while before you begin to pour the wax into the molds, put the clean wax container on the second hot plate to keep it warm.

While the wax is warming, prepare your mold and wick. Most polyurethane molds don't need a spray of mold release—you'll learn if yours does or not. (Check the instructions that came with your mold.) String the wick up through the bottom of the mold. You may need to push it through with a nail, wire, or hooked needle made especially for this purpose. Pull up the wick and attach it to a support. A large bobby pin works well. Secure the wick in the very center and pull it taut. Wax probably won't leak, but protect the surface the first time if you're not sure. The opening of the mold will be the unseen bottom of the candle, so the finish isn't critical, but it can be easily repaired if it turns out uneven or unsightly.

When the cleaned wax is just melted, it's cool enough to pour. First place the mold on a small tray or other covering. Slowly fill the mold, avoiding splashes and bubbles in the wax. The slower the wax cools, the less likely the candle is to be deformed or to crack. If, after the wax has cooled, the bottom recedes, gently refill the mold and cool again.

Pour the melted wax through the final filter, shown here secured with clothespins. Protect your surface, and place the container with the clean wax on a hot plate switched to the warm setting, if needed.

🐝 *When the mold and wick are ready, fill the mold with barely melted wax. Top off the mold with more melted wax if the bottom shrinks after cooling. After the wax is completely cooled, carefully remove the candle from the mold.*

🐝 *You can easily make seasonal ornaments with embedded hangers. Use decorative string or ribbon for the hanger. Fill a decorative candy mold about halfway, lay the hanger in the wax, and finish filling. The loose ends of the hanger are hidden and secure.*

When completely cool—about a day at average room temperature—untie the wick and carefully remove the candle. Don't be in a hurry, especially if you are using an intricate mold with lots of detail. Remove any rubber bands used to hold the mold together, and gently pull back the sides of the mold, exposing the cooled candle.

Lift up the candle and pull enough wick into the mold to make another candle—this trick eliminates the wick-threading task next time.

If the bottom of the candle is uneven, you can smooth it out. Warm a small pan with aluminum foil on the bottom on the hot plate and rub the bottom of the candle until smooth and even. Pour the melted wax back into your container. Your candle is finished and ready to burn.

Before you start sharing these wonderful gifts with friends and family, or even selling them, make sure they work. You'll have to test-burn a couple of each kind you make to make sure the wax is clean and you have the right-sized wick. You don't need to burn them completely if wax is in short supply, but use at least half when testing a small candle.

If the candle burns with a steady, clean flame, and without smoke, drips, or sputtering, you can be fairly certain you've chosen all the right ingredients. But if it doesn't burn well, figure out why: it's either dirty wax, a poorly sized wick, or air bubbles in the wax. Make adjustments, such as cleaning the wax again or choosing a larger or smaller wick. Melt the test candle to use again.

Candle Safety

> NEVER LEAVE A BURNING CANDLE UNATTENDED.

> PLACE CANDLES OUT OF REACH OF CHILDREN AND PETS.

> DISCARD (OR REMELT) A CANDLE STUMP WHEN IT REACHES ABOUT 1" (2.5 CM) IN LENGTH.

Making Cosmetic Creams

Beeswax is an essential ingredient in many cosmetics. It is used because it is an all-natural, hypoallergenic substance that adds body and texture to the finished product, has a delicate aroma, and is a wonderful skin softener and protector. It is used in hand, foot, and body creams; lipstick; eye shadow; lip ointment; and fine soaps.

Most of these creams are easy to make, even in the kitchen, and are a useful way to use small amounts of beeswax. The other ingredients are available at health stores, as well as many sources available on the Internet.

Included here are simple recipes for using your beeswax in some homemade beauty products. If you're going to make these as gifts, many attractive molds and ornamental containers are available from the same sources as the ingredients. You can make nearly any shape or find nearly any kind of container. Plus, many how-to resources—books, videos, and Web pages—are available for making beauty products. Experiment with the recipes provided here, and enjoy yet another contribution from your bees.

Jeanne's Hand Cream

Jeanne Schell keeps bees in Medina County, Ohio, with the help of her daughter, Katie. Jeanne says the beauty of her simple hand cream recipe is that you can substitute varying amounts of shea butter or cocoa butter for the palm oil and coconut oil if you want to customize the aroma and thickness of your cream. Experiment with varying amounts of oils and butters to get just the right mix for your hands. Just make sure that the combined amount of oils and butters, other than base oil (olive oil), equals a full cup. You can also vary the amounts of the ingredients to make more or less of this cream, but the basic recipe fills about one dozen 2-ounce (55 g) containers—more than enough for your use, with lots of extras to share.

This easy-to-make hand cream for everyday use will protect and soothe your hands after the rigors of home and garden tasks.

2 cups (475 ml) olive oil (the base oil)

¼ cup (60 ml) palm oil

¾ cup (175 ml) coconut oil

6 ounces (170 g) beeswax

40–50 drops essential oil(s) (optional)

Combine the oils in a 2-quart (1.9-l) stainless-steel sauce pan, and stir over medium heat until the oils (and butters, if using) are melted. Add the beeswax to the pan, and stir until melted. Then test the mix by dropping five or six drops onto a sheet of waxed paper, let cool, and test the hardness. If it is too hard, it will be difficult to rub onto your skin, and your mix will need a bit more base oil added to soften it. If it is too soft or greasy, add a bit more beeswax to stiffen it.

When the hand cream has the exact thickness you want, remove the pan from the burner and allow it to cool until the cream begins to harden on the sides of the pan. Then stir in two vitamin E tablets or six drops of vitamin E oil to enhance the healing properties of the hand cream.

For a fragrant hand cream, add any combination of essential oils after removing the pan from heat, or simply rely on the subtle fragrance of beeswax and the oils in the basic recipe. A popular blend that offers a whiff of fragrance without being overpowering contains 15 drops of lavender oil, 15 drops of rosemary oil, and 15 drops of geranium oil. You can use other oils, or vary the proportions of these, to suit your taste.

When well mixed, pour the cooling cream into individual 2-ounce (55 g) containers and allow it to cool, uncovered, overnight. Then cover the containers, label them, and store at room temperature, out of direct sunlight.

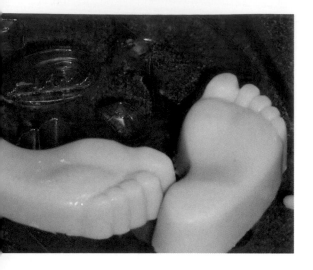

Jeanne's Famous Favorite Foot Cream

Jeanne and her daughter Katie make creams and soaps using beeswax and other products. They've been at it for some time and have simplified the process quite a bit. They make and sell their products at craft fairs, farmer's markets, and other places. This is Jeanne's favorite foot cream recipe.

scale that weighs in ounces or grams

stainless steel 2-quart (1.9 l) pot

plastic stirring spoon

molds

mold release

2 ounces (55 g) beeswax (small broken pieces)

4 ounces (113 g) shea butter

peppermint essential oil, 20–30 drops
   (lighter to stronger aroma)

2 ounces (55 g) deodorized cocoa butter

Shea butter is a pale, solid fat pressed from the seeds of the shea tree, commonly used in foods, creams, soap, and even candles. The unrefined product is normally used in creams and has a hint of chocolate aroma. Add to that the chocolate aroma of cocoa butter, the fragrance of beeswax, and a hint of peppermint, and you have a very pleasing mix.

Melt cleaned beeswax and 2 ounces (55 g) of cocoa butter in the stainless steel pan until they are just barely melted. Use only low to moderate heat. When ready for the next step, the mix will be just barely clear.

Add 4 ounces (113 g) of shea butter and stir, allowing it to melt in the beeswax-cocoa butter mix. Reduce the burner temperature and let the mix slowly cool until it reaches the can-just-barely-pour stage—a thermometer isn't needed here.

At that stage, add the peppermint oil. You want the mix as cool as possible so it doesn't volatilize the fragrance of the peppermint oil.

Spray your molds with mold release while the mix is cooling, and set the molds on a protected flat, solid surface. A kitchen counter is ideal, but this takes some space while cooling. If you need to move them when finished, place them on a board beforehand so they don't tip or squeeze. When ready, fill the molds full. Let cool overnight. If the bottom is uneven, level using the hot pan technique for candles, but not so hot it melts too fast. When cool, remove and place in a package that seals the bar—for home, a sealable plastic bag; for sales, a clean plastic wrap, plus ribbons and bows—so the fragrance doesn't escape too fast.

Use Jeanne's foot cream for rough and dry spots on hands, feet, elbows, and knees. It is fragrant and pleasant to use… and smell.

When all done, the pan and molds can be washed up in hot dishwater without fear of clogging drains or ruining pots or spoons. Spills, too, clean up well with soap and water. From start to finish this project, from gathering your materials to setting aside the filled molds, takes less than a half hour. Beeswax creams are fast and easy to make, useful for friends and family, and even profitable.

Weigh out 2 ounces (55 g) of beeswax, and 2 ounces (55 g) of cocoa butter. Put them in a stainless steel pan and melt them together over moderate heat.

When the beeswax and cocoa butter are just barely melted, add the shea butter to the mix, stir until melted, and reduce the heat. While cooling the mixture, spray the molds with mold release. When the mix cools to the point where you can barely pour it, stir in the peppermint oil.

When well mixed, pour the liquid into the molds slowly and carefully, as close to the mold as reasonable to avoid splashing and bubbles in the cream later.

Let the cream set in the molds for 24 hours at room temperature to cool and harden.

When cool, remove the bars from the molds carefully so they aren't damaged. These types of creams are soft.

Store in a sealable plastic bag to protect the volatiles and to keep the cream moist. Use sparingly—a little goes a long way

Making Soap

Making a beeswax-based soap is another way to use small amounts of this fine material. Soap containing beeswax—often with essential oil fragrances and natural, earthen dyes—makes a unique gift and is especially nice in the bath.

Making soap is a bit more complicated than making creams and requires care and attention to detail, but with practice and patience, you will soon be making different varieties, colors, fragrances, and shapes.

The recipe included here is a basic beeswax formulation, scented with natural oils and using natural ingredients. The entire process, from gathering ingredients to covering the trays, takes about two hours. It will take more time for the first few batches, but not much less, even with practice.

✿✿✿✿✿✿✿✿✿✿✿✿✿✿✿✿✿✿✿✿✿

Start by assembling all your equipment and ingredients.

Tools

1 tray mold, sprayed well with nonstick coating

board(s) to cover tray and to hold finished mold

blanket to insulate while setting up (two to three days required)

measuring cup marked in ounces or grams

gallon (3.8 l) plastic container

scale weighing half ounces or grams

nonreactive quart pan (such as enamel or stainless steel)

ice cubes

nonreactive stir sticks, wood or plastic

dishwashing gloves

protective eyewear

candy thermometer

sharp knife

Ingredients

32 ounces (907 g) coconut oil, solid form

28 ounces (794 g) solid vegetable oil (Crisco works well)

4 ounces (113 g) beeswax, shaved or very small pieces

10¼ ounces (298 g) lye (NaOH sodium hydroxide)

24 ounces (710 ml) distilled water

2 ounces (60 ml) essential oil blend

1¼ ounces (40 ml) sweet orange oil

½ ounce (14 ml) Litsea oil (lemony aroma)

½ ounce (14 ml) ylang-ylang oil (jasmine scent)

2 level teaspoons red ochre dye

✿✿✿✿✿✿✿✿✿✿✿✿✿✿✿✿✿✿✿✿✿

Note:
WEAR GLOVES AND EYE PROTECTION AT
ALL TIMES. LYE WATER IS CORROSIVE AND
WILL CAUSE SKIN BURNS ON CONTACT.

Begin by weighing all the ingredients—water, lye, oils, fragrances, and dyes—before you begin. Put the water required, 24 ounces (680 g), in the large plastic 1-gallon (3.8 l) container. Fill a large pot or stainless steel sink with twenty to fifty ice cubes and 3" or 4" (7.6 or 10.2 cm) of cold water.

Weigh the lye (NaOH drain cleaner) in a dry plastic measuring cup.

Add the lye to the water, *not* water to the lye, and stir constantly but gently until it dissolves. *Do not splash!* The water mixture will become very hot as you stir, reaching nearly the boiling point. Place the container in the ice bath to cool, stirring occasionally.

Place the enamel or stainless steel pan on the stove with moderate heat. Add the coconut oil, vegetable oil, and beeswax. Stir regularly until all ingredients melt, and stir until everything is completely blended. Use the thermometer to keep tabs on the temperature and stir until the temperature reaches 110° F to 120° F (43° C to 49° C).

Don't forget about the water/lye mixture cooling in the sink during all this.

When the lye mix has cooled enough that it is warm to the touch, and the oil mix is also at the right temperature, it's time to combine them.

Turn off the stove and remove the pan from the burner. Very slowly add the lye water to the oil mix, stirring constantly. This step takes a long time, so be patient. The mix will warm again, and you need to stir until it cools and all the lye water has been added. Allow at least 15 minutes for this step.

When the lye water has been added, slowly stir in the dye, mixing thoroughly. As the mix cools and the dye is completely blended, the mixture will begin to thicken, and you'll begin to see a trail, or *tracing* as it's called, behind the spoon as your stir. When this occurs, add your fragrance. You want the mix as cool as possible, so that the fragrance volatilizes as little as possible. Stir well so it is completely mixed in.

Make certain that your mold is level and on a solid, covered, and safe surface. Pour the mix until the mold is completely full. Let the surface dry for a few moments, then cover with cardboard or wooden board, and top that with a blanket, for insulation. Let the mold cool overnight at room temperature. In 24 hours, turn the soap over, making sure the soap is supported by the board, and remove it from the mold. Cover it with the blanket. Turn soap again in 24 hours, and once more 24 hours after that.

When the soap has cured for three days, turn it onto a safe surface, and cut to size using a large, sharp knife. Set pieces of soap on a rack or board, and let cure in the open air for another thirty days. The individual bars will shrink a bit during the airing process.

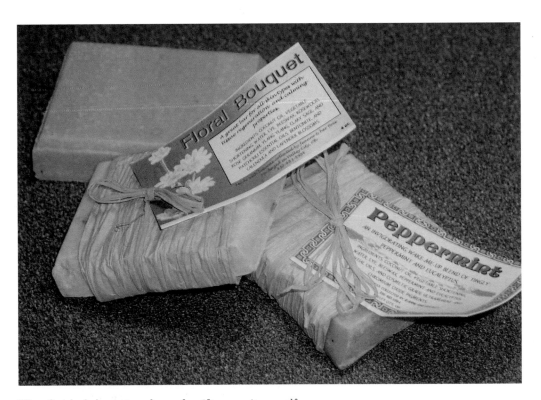

When finished, decorate and wrap for gifts or use it yourself.

Other Beauty Benefits from Your Hive and Garden

Containers of lip balm made from honey, beeswax, and other natural ingredients offer a fragrant cure for chapped lips.

Nancy's Lip Balm

Nancy Riopelle makes and sells lip balm from the beeswax and honey that her husband Buzz (yes, that's his name) produces from his 100 hives each season. She sells lots of this recipe at their local fairs and farm markets in Medina County, Ohio, each summer and fall.

This recipe makes enough lip balm to fill 100 0.15-ounce (4 g) lip-balm tubes, or 65 ⅓-ounce (9 g) pots.

1 cup (225 g) shredded beeswax

14 oz. (425 ml) coconut oil

5 tbsp. (100 g) honey

5 tbsp. (70 ml) pure vanilla extract

Heat the wax in a saucepan over low heat to 150° F (66° C). In a separate saucepan, heat the oil to the same temperature. When both are heated to the proper temperature, add the coconut oil to the beeswax, remove the pan from heat, and stir steadily until well blended. Then add the honey and the vanilla extract, and continue to stir until well blended. Pour into tubes or tubs, allow to cool overnight, and then cap the containers and store at room temperature, out of direct sunlight.

Honey Cucumber Skin Toner

This refreshing skin toner has only two ingredients: one from the garden and one from the hive.

medium size cucumber

2 tsp. (20 g) honey

Peel and cut the cucumber into pieces, and put the pieces into a blender and make a puree. Take a colander and set atop a small bowl. Line the colander with the same material you used to strain honey, and place the cucumber puree in the colander, letting the juice drain through the filter material into the bowl. When completely drained, add the honey to the liquid and mix until the honey is completely dissolved. Pour into a sealable bottle and store in the refrigerator.

To use, shake the bottle to remix, and soak a cotton pad with the toner. Briskly brush over your face and neck first thing in the morning (after your shower), and just before bed at night, allowing it to dry. This mixture lasts about a week.

Uses for Beeswax around the House

Pure beeswax has a multitude of uses around the home. Rub on sticky or squeaky drawer edges for smoother movement. Put on screw threads or nails for easier entry and waterproofing. Draw sewing needles across a block of beeswax to make them easier to pull through fabric. You can purchase small molds that have the word *beeswax* embossed on the top, and then sell these 1-ounce (28 g) blocks of beeswax or give away as gifts. And, you can simply melt beeswax into chunks to sell to other beekeepers.

CHAPTER 5 →About Honey←

granulated honey (also called crystallized). Freezers, however, retard this process and are perfect for long-term storage.

To Liquefy Granulated Honey

All honey will granulate eventually—some in weeks, others in years. This process is natural and in no way affects the quality or flavor of the honey. You can return honey to a liquid state simply by warming it. But remember, honey and high heat do not mix. The delicate and volatile flavors of honey can be easily driven off—or worse, damaged—by overheating.

If your honey has granulated, and you want it to return to a liquid, place the bottle in a pan of warm (not hot) water on the stove. Keep the burner on the lowest possible setting, loosen the cap, and patiently wait for the honey to liquefy.

Or, remove the cover and place the jar in a microwave oven. Set the microwave on medium for one or two minutes and see how much becomes liquid. Repeat, with two to three minute intervals between, until complete. Do not overheat.

Honey should not be overheated. Water-bath temperatures must not exceed 100° F (38° C). Extreme care should be taken when heating any glass bottle to avoid cracking the glass. Heating to temperatures over 120° F (49° C) can lead to this.

Cooking with Honey

To measure honey without hassle, dip your spoons or cups in a little oil so the honey simply slips out. Warming these utensils helps, too.

Honey turns brown and burns easily when cooked, and special care should be used when baking or heating honey in any way. If possible, reduce oven temperature a bit to lessen the risk of overheating.

Dark honey varieties are more sensitive to overheating than lighter varieties, primarily because of the greater amount of minerals, vitamins, and proteins present.

Recipes with Honey and Your Garden Harvest

The recipes that follow include your harvested honey as well as contributions from your garden or orchard. The bees pollinate, Mother Nature grows, you combine the final products, and your family and friends enjoy. The perfect culmination of bees working in your garden!

There are only a few things you need to know when baking or cooking with nature's natural sweetener. First and foremost, honey is the perfume of the flowers from which it was taken. It gently holds the sunshine from flower-filled meadows, the natural sweetness of a warm spring morning, and the tranquility of a summer rain. Honey is a delicate and subtle sweetener, whose infinite flavors and colors make every variety a wonderful and exciting experience to taste.

To use honey effectively and tastefully, simply follow these hints and suggestions gathered from honey users over hundreds of years and thousands of recipes.

Using Honey

To keep liquid honey in its liquid state, store it in a warm, dry location. The kitchen table is the perfect place for it; refrigerators are not. Cool temperatures—below 60° F (16° C)—result in

meats

Salsa Salmon Fillets

4 large salmon fillets
2 tbsp. (40 g) honey
1 cup (150 g) tomatoes, thinly sliced
1 green onion, sliced thin
¼ cup (30 g) red onion, finely chopped
2 tbsp. (28 ml) white wine vinegar
2 tbsp. (28 ml) fresh lime juice
2 tbsp. cilantro, chopped
¼ tsp. salt
1 jalapeño pepper, seeded and chopped
1 cup (130 g) frozen corn
Vegetable spray

Mix together all ingredients, except for the salmon, honey, and corn, and set aside.

Sauté the corn until heated in a large, nonstick frying pan coated with vegetable spray. Stir the corn into the salsa that you already have mixed.

To the same pan, add salmon fillets and sauté until done. Drizzle the honey over the fillets while they are cooking. Cook 4–5 minutes on each side until the fillets flake.

Put a fillet on a plate and add a tablespoon or two of salsa on each fillet. As a side, add fresh melon or fruit slices.

Honey Lamb Chop Glaze

¾ cup (220 g) honey, light and strong
½ cup (120 g) prepared mustard
⅛ tsp. onion salt
⅛ tsp. lemon pepper
8 lamb chops, at least 1" (2.5 cm) thick

Heat honey, mustard, onion salt, and lemon pepper, stirring occasionally. Keep warm over low heat.

Broil chops about seven minutes, and then brush them with the glaze. Turn the chops; broil five minutes longer, brushing again with honey mixture. Serve remaining glaze with chops.

Do-It-Yourself Honey-Baked Ham

5-lb. (2.3 kg) ready-to-eat ham
¼ cup (40 g) whole cloves
¼ cup (85 g) dark corn syrup
2 cups (680 g) honey
⅔ cup (150 g) butter

Preheat oven to 325° F (163° C). Score the ham and stud it with cloves. Place the ham in a foil-lined pan. In the top of a double boiler, warm dark corn syrup, honey, and butter. Brush this glaze over the ham and bake for 75 minutes, basting every 10–15 minutes with warm glaze. During the last 4–5 minutes of cooking time, turn on the broiler to caramelize the glaze. Remove from the oven and let sit a few minutes before serving.

Honey Baked Chicken

¼ cup (85 g) honey
2 tbsp. (28 ml) balsamic vinegar
1¼ cups (205 g) dried bread crumbs
1 tbsp. (15 ml) olive oil
6 boneless, skinless chicken breast halves, about 2 lbs. (.9 kg)

In a shallow bowl, whisk together honey and vinegar. Pour bread crumbs into a separate bowl. Set both bowls aside. Spread the oil over the bottom of a foil-lined baking pan large enough to hold all chicken pieces in one layer. Roll the chicken pieces in the honey and vinegar, and then roll them in bread crumbs, and place them in the pan. Bake at 375° F (191° C) for 30 minutes, or until cooked through. Makes 4 servings.

Southern-Style Honey Barbecued Chicken

1 2½–3 lb. (1.1–1.4 kg) chicken quartered
1 cup (160 g) onions, thinly sliced
¾ cup (180 g) tomato sauce
¼ cup (85 g) honey
¼ cup (60 ml) cider vinegar
2 tbsp. (28 ml) Worcestershire sauce
1 tsp. paprika

Place chicken, skin side down, in single layer in large baking dish. Sprinkle with salt and pepper. Combine remaining ingredients; mix well. Pour mixture over chicken. Bake, uncovered, at 375° F (191° C) for 30 minutes; turn the pieces and bake 20 minutes more, or until the chicken is glazed and thoroughly cooked. Makes four servings.

TOPPINGS

Honey-and-Spice Blueberry Syrup

1½ cups (660 g) honey, clover or other mild flavor
½ cup (120 ml) water
½ tsp. ground cinnamon
1¾ cups (260 g) fresh blueberries
1 tbsp. (14 ml) lemon juice
½ tsp. (3 ml) vanilla

Put honey, water, and cinnamon in a large saucepan and bring to a boil. Reduce heat to low and simmer 10 minutes, stirring until sauce thickens. Allow the syrup to cool until it is just warm. Add remaining ingredients and stir. Use as a syrup for waffles, pancakes, French toast, ice cream, or pound cake. Makes 2⅔ cups.

Honey Berry Puree

2 cups (300 g) ripe blackberries or raspberries
½ cup (170 g) honey
2 tbsp. (28 ml) brandy

Purée berries in blender or food processor, and press through sieve to remove seeds, or mash berries with fork. Stir in honey and brandy, until blended. Serve over angel food cake, sliced fruit, or ice cream. Makes two cups.

Honey Chocolate Sauce

1½ cups (510 g) honey
1½ cups (150 g) unsweetened cocoa powder
2 tbsp. (28 g) butter

Combine all ingredients in small bowl; mix well. Cover with plastic wrap and microwave at high (100%) 2 to 2½ minutes, stirring after one minute. Pour into sterilized gift jars. Keep refrigerated.

Orange-Honey Butter for Corn Bread

½ cup (112 g) unsalted butter, at room temperature
⅛ tsp. salt, or to taste
1 tbsp. orange zest, finely grated (1 medium to large orange)
1 tbsp. (20 g) honey

Put the softened butter into a bowl with the salt and whisk until creamy. Whisk in the orange zest and then the honey. Whisk until smooth.

Warm corn bread at 250° F (121° C) for 5–10 minutes. Remove from the oven and brush with a little orange-honey butter. Cool about 15 minutes before cutting into wedges. Serve with the remaining butter.

Orange Cream Spread

1 8-ounce (227-g) package cream cheese
¼ cup (170 g) honey, mild
2 tbsp. (28 ml) orange juice
½ tsp. orange peel

Combine softened cream cheese, honey, orange juice, and orange peel; blend well. Refrigerate at least one hour—overnight is better. Spread on rolls, muffins, or croissants.

Peanut Honey Spread

¾ cup (225 g) peanut butter
½ cup (170 g) honey
1 tsp. ground cinnamon

Combine ingredients and mix thoroughly. Spread on English muffins, biscuits, or sandwiches.

Heavenly Chocolate Honey Dip

1 cup (230 g) sour cream
½ cup (170 g) honey, medium flavor
¼ cup (50 g) unsweetened cocoa powder
1 tsp. (5 ml) vanilla

Combine all ingredients in medium bowl and blend thoroughly. Cover and refrigerate until ready to serve. Serve with assorted fruits or chunks of angel food cake. Makes six servings.

Honey Rhubarb Compote

⅔ cup (225 g) honey
1 cup (235 ml) water
4 cups (400 g) rhubarb, coarsely chopped into ½" (1.3 cm) pieces
½ tsp. (3 ml) vanilla
2 tbsp. cornstarch
3 tbsp. (45 ml) cold water
2 pints (570 g) frozen yogurt

Dissolve honey in water in large, nonaluminum saucepan. Bring to a boil over medium-high heat. Add rhubarb. Reduce heat to low and simmer uncovered, 15 to 25 minutes or until rhubarb is tender but still intact. Stir in vanilla. Combine cornstarch with 3 tablespoons (45 ml) of water and mix well. Gradually stir cornstarch mixture into rhubarb and cook. Stir until mixture comes to a boil. Reduce heat; simmer for 3–5 minutes, or until the mixture thickens. Pour into serving bowl and refrigerate until cold. Serve over frozen yogurt or ice cream.

Blueberry-Studded Honey Peach Sauce

1 cup (340 g) honey
¼ cup (38 g) fresh blueberries
1 tsp. ground cinnamon
1 quart (600 g) fresh peaches, sliced

Combine honey, blueberries, and cinnamon in large saucepan, mix well and bring to a boil over medium-high heat. Reduce heat to low; simmer 10 minutes or until flavors are blended. Remove from heat. Add the peaches and mix. Serve over waffles, pancakes, or ice cream.

Whipped Honey Butter

1 cup (225 g) whipped or creamed honey, softened
½ lb. (225 g) butter, softened
1½ tsp. orange peel, grated

In a medium bowl, mix together honey and butter. When mixed, add orange peel. Spoon into jars with tight-fitting lids. Store in refrigerator.

SIMPLY SWEET IDEAS

Sweet Touch

Mix honey with melted butter. Drizzle over angel food cake or pound cake for a quick finishing touch.

No-Fuss Frosting

To frost carrot cake and cupcakes, soften an 8-ounce (227-g) package of cream cheese and mix with ¼ cup (85 g) of light, mild honey and a pinch of salt. Cream until smooth.

fruit recipes

Apples, Waffles, Ham, and Honey

¾ cup (255 g) honey, divided (in two separate cups)
¼ cup (60 ml) apple juice
2 tbsp. (28 g) butter
2 crisp red apples, cored and sliced
8 frozen waffles, toasted
8 slices deli ham

In a small saucepan, combine ½ cup (170g) honey and apple juice. Stir over medium heat until heated through. Set aside and keep warm. In large nonstick skillet, melt butter with remaining ¼ cup (85 g) honey. Add apples; cook and stir until apple slices are lightly caramelized and crisp-tender. Place one waffle on each plate. Top each with two ham slices and another waffle. Top each with apple slices and syrup. Makes four servings.

Honey-Caramelized Bananas and Oranges

2 large bananas, peeled and halved lengthwise
1 orange, peeled and sliced
¼ cup (85 g) honey
3 tbsp. (20 g) chopped walnuts
3 tbsp. (45 ml) brandy

Place bananas and orange slices in small flameproof dish. Drizzle with honey and sprinkle with walnuts. Broil 4–6 inches from heat source for about 5 minutes, or until heated through and lightly browned. Remove from broiler. Pour brandy over top and flame. Serve immediately. Makes two servings.

Brownie Sundaes with Honey-Berry Sauce

¼ cup (85 g) honey
1 tbsp. (15 ml) lemon juice
½ tsp. lemon peel, grated
2 cups (450 g) raspberries, blackberries, or strawberries
4 chocolate brownies, purchased or prepared
1 pint (285 g) vanilla ice cream

In small bowl, combine honey, lemon juice, and lemon peel. Mix well. Gently mix in berries. Place brownies on serving plate and top with a scoop of ice cream. Drizzle with ¼ cup (120 ml) sauce. The sauce is also especially good served over grilled pound cake, or in crepes. Makes four servings.

Fruit Salad with Honey-Orange Dressing

½ cup (125 g) plain yogurt
¼ cup (55g) mayonnaise
¼ cup (85 g) honey
¾ tsp. orange peel, grated
¼ tsp. dry mustard
3 tbsp. (45 ml) orange juice
1½ tsp. (20 ml) vinegar
4 cups (600 g) assorted fruit—berries, melon balls, grapes, apple, or pear slices

Whisk together yogurt, mayonnaise, honey, orange peel, and mustard in small bowl until blended. Gradually mix in orange juice and vinegar. Toss fruit gently with dressing. Cover and refrigerate until ready to serve. Makes four servings.

drinks

Hot Spiced Tea

4 cups (940 ml) freshly brewed tea
¼ cup (85 g) honey
4 cinnamon sticks
4 whole cloves
4 slices mandarin orange

Combine tea, honey, cinnamon sticks, and cloves in medium saucepan; simmer for 5 minutes. Strain into mugs. Garnish with orange slices and serve hot. Makes four servings.

Honey-Lemonade Fruit Cubes

1 ½ (180 ml) cups lemon juice
¾ (225 g) cup honey
9 cups (2.1 Liters) water
48 small pieces of assorted fresh fruit

Combine lemon juice and honey in large pitcher, and stir until honey is dissolved. Add water and stir. Place one to two pieces of fruit in each compartment of two ice cube trays, then fill each with the lemonade and freeze until firm. Chill remaining lemonade.

To serve, put frozen fruit cubes in tall glasses and fill with remaining lemonade. Makes nine cups.

Honey-Berry Milkshakes

1 pint (285 g) vanilla ice cream
2 ½ cups (425 g) strawberries or assorted berries (raspberries, blueberries, blackberries)
½ cup (120 ml) milk
¼ cup (85 g) honey, light
4 small mint sprigs

Put everything except the mint in a blender or food processor, and process about 30 seconds or until smooth. Pour into tall glasses. Garnish with mint sprigs. Makes four cups.

Cool Green Tea

1 pint (473 ml) fresh strawberries, hulled
¼ cup (85 g) honey, light and mild
1 6-ounce (210-g) can frozen orange juice concentrate
2 cups (470 ml) green tea, cooled

In a blender, combine strawberries and honey, and blend until smooth. Add orange juice concentrate and blend again. Stir this mixture with the cooled tea and serve over ice.

Honey-Berry Float

1 quart (9464 ml) milk
6 tbsp. (120 g) honey
2 cups (340 g) crushed fresh strawberries
¼ tsp. (3 ml) almond extract
1 quart (570 g) vanilla ice cream
sprigs of apple mint

Combine milk, honey, strawberries, almond extract, and one pint (285 g) ice cream. Blend briefly until smooth. Pour into tall glasses and garnish with scoops of ice cream and a sprig or two of apple mint.

Cool Latte

2 cups (470 ml) double-strength brewed coffee
1 cup (235 ml) milk
¼ cup (85 g) honey
2 cups (300 g) ice

In a large pitcher, stir together the coffee, milk, and honey until thoroughly combined and honey is dissolved. Chill. Just before serving, blend this mixture with ice in blender until frothy and smooth.

veGeTaBLes

Pumpkin Soup

1 cup (130 g) yellow onion, chopped
2 medium cloves garlic, minced or pressed
1 tsp. pumpkin pie spice
3 ¾ cups (880 ml) chicken broth
15-ounce (425 g) can pumpkin puree
1 tbsp. (20 g) honey
1 cup (235 ml) milk
4 tsp. cornstarch
1 tbsp. (14 ml) fresh squeezed lemon juice
dash sea salt
2 tbsp. (15 g) sour cream or plain yogurt

In a 5-quart saucepan, stir onion, garlic, spices, and ¼ cup broth over high heat until pan is dry, about 2–3 minutes. In a food processor, puree onion mixture until smooth, adding a little more broth if needed. Return puree to pan; add remaining broth, pumpkin, and honey. Mix milk with cornstarch until smooth; stir into soup. Bring to a boil over high heat, stirring constantly. Add lemon juice, then remove from heat.

To serve, ladle soup into bowls. Stir sour cream to soften, then drizzle a small dollop into center of each bowl. Makes six servings.

Honey Carrots

3 cups (390 g) carrots, sliced
¼ cup (85 g) honey
2 tbsp. (28 g) butter
2 tbsp. fresh parsley, chopped

Heat 2" (5 cm) salted water in medium saucepan to a boil over high heat. Add carrots and return to a boil. Reduce heat to medium-high. Cover and cook 8–12 minutes, until carrots are crisp-tender. Drain carrots; return to saucepan. Stir in honey, butter, and parsley. Cook and stir over low heat until carrots are glazed. Makes four servings.

Mixed Greens with Shrimp

¼ cup (85 g) honey
¼ cup (60 ml) white wine vinegar
8 cups (160 g) mixed greens
½ pound (225 g) cooked shrimp, shelled and deveined
2 tbsp. minced fresh mint
salt and pepper to taste

In small bowl, whisk together honey and vinegar. Set aside. In large bowl, combine greens, shrimp, and mint. Gently toss salad with dressing. Season with salt and pepper to serve. Makes four servings.

Garden Leftovers

6 zucchini (or other summer squash)
2 tbsp. (28 g) butter
¼ cup (85 g) honey
2 cups (450 g) tomatoes
1 cup (115 g) bread crumbs

Slice zucchini lengthwise, and cut into 2"–3" (5–7.5-cm) pieces. Sauté lightly in butter. Add honey. Cook and strain tomatoes. Bring the entire mix to a boil. Cook 5 minutes longer. Add the bread crumbs with tomatoes, if desired, to thicken the finished mix. Serves three to five.

Asparagus with Honey-Garlic Sauce

1 lb. (.5 kg) fresh asparagus
¼ cup (60 ml) Dijon mustard
¼ cup (60 ml) dark ale
3 tbsp. (60 g) honey
½ tsp. garlic, minced
¼ tsp. dried thyme leaves, crushed
¼ tsp. salt

Add asparagus to boiling, salted water, cover and cook, about 2 minutes or until tender. Remove and drain. Combine mustard, ale, honey, garlic, thyme, and salt; mix well, and pour over the asparagus, just before serving. Makes four servings.

Summer Squash Surprise

4 summer squash
4 tbsp. (55 g) butter
⅓ cup (115 g) honey
dash salt
¼ tsp. cinnamon
dash nutmeg

Combine all ingredients except squash. Slice down the middle of each squash (zucchini, yellow summer, crookneck, or scallop) and scoop out pulp. Add blended ingredients to the pulp, mix thoroughly and replace in squash shells. Place a rack in a shallow baking dish, add water to ¼" (.6 cm). Place squash on rack. Bake at 350° F (177° C) for 25–30 minutes. Makes six to eight servings.

salad dressing

Honey Dill Dressing over Red-Skinned Potatoes

1 ½ lbs. (.7 kg) small new red potatoes
4 strips bacon
1 medium red onion, diced
6 tbsp. (120 g) honey
6 tbsp. (90 ml) apple cider vinegar
½ tsp. cornstarch
½ tsp. water
2 tbsp. chopped fresh dill
1 bunch watercress, chopped

In large pot, boil the potatoes whole with skins on in salted water until tender but firm. Drain and cool. While potatoes cool, sauté bacon until crisp, then set aside. Add onion to pan with bacon drippings and cook until soft. Add honey and vinegar to the pan, stir and bring to a boil. Blend cornstarch with water and stir into the honey mixture. Cook until mixture thickens, then remove from heat. Next, crumble the bacon and stir it and the dill into dressing. Cut now-cooled potatoes in half or quarters. Put the potatoes and watercress in a large bowl, pour dressing over, and toss gently.

Tangy Lemon Dressing

¼ cup (60 ml) olive oil
½ tsp. lemon peel, grated
¼ cup (60 ml) lemon juice
¼ cup (60 ml) water
2 tbsp. fresh chives, chopped
1 tbsp. (15 ml) Dijon mustard
1 tbsp. (20 g) honey
¼ tsp. salt

Shake all ingredients in tightly covered container. Makes about ¾ cup dressing.

Classic Salad Dressing

½ cup (170 g) honey
½ cup (120 ml) vinegar
½ cup (120 ml) canola or olive oil
1 tbsp. (40 g) onion, chopped
1 tbsp. (40 g) green pepper, chopped
2 tbsp. celery seed
dash salt

Blend ingredients well. Chill before serving.

sauces

BBQ Sauce

1 onion, diced
1 tbsp. ground coriander
2 tbsp. (40 g) honey, strong
¼ tsp. pepper
¼ tsp. garlic powder
⅛ tsp. cayenne pepper
3 tbsp. (45 ml) lemon juice
¼ cup (60 ml) soy sauce

Combine all ingredients in saucepan and simmer for 5 minutes. When grilling, begin applying barbecue sauce only after meat is half-cooked to prevent burning the sauce.

Barbecue Baste

1 tbsp. (15 ml) vegetable oil
¼ cup (30 g) onion, minced
1 clove garlic, minced
½ cup (125 g) tomato sauce
⅓ cup (113 g) honey
3 tbsp. (45 ml) vinegar
1 tsp. dry mustard
½ tsp. salt
¼ tsp. black pepper

Heat oil in saucepan over medium heat until hot. Add the onion and garlic and cook, stirring until onion is tender. Add the rest of the ingredients and bring to a boil; reduce heat to low, and simmer about 20 minutes. Serve over grilled chicken, pork, spareribs, or hamburgers. Makes one cup (235 ml) .

Smokey Honey Barbecue Sauce

1 cup (340 g) honey
1 cup (225 g) chili sauce
½ cup (120 ml) cider vinegar
1 tsp. (5 ml) prepared mustard
1 tsp. (5 ml) Worcestershire sauce
¼ tsp. pepper
½ tsp. minced garlic
2 or 3 drops smoke flavoring

Combine all ingredients except smoke flavoring in medium saucepan over medium heat. Stir frequently for 20–30 minutes. Remove from heat and add smoke flavoring to taste. Serve over grilled chicken, turkey, pork, spareribs, or hamburgers. Makes about two cups (475 ml).

Honey Vegetable Sauce

¼ cup (85 g) honey
¼ cup (56 g) butter
2 tbsp. (20 g) onion, minced
½ tsp. fresh thyme, crushed
salt and pepper to taste

In saucepan, combine all ingredients and bring to boil; cook for 2 minutes. Makes about ½ cup (120 ml) . Toss with cooked cauliflower or carrots, or mix with baked squash.

salads

Apple Salad

2 cups (300 g) apple, chopped

1 cup (150 g) celery, chopped

½ cup (75 g) dates, chopped

4 tbsp. (60 ml) fresh lemon juice

2 cups (300 g) seedless grapes, cut in halves or quarters

¼ cup (85 g) honey

½ cup (120 g) mayonnaise

¼ tsp. salt

2 tbsp. (28 ml) heavy cream

½ cup (63 g) chopped nuts (walnuts or sunflower seeds are best)

Combine apples, celery, dates, grapes, and lemon juice in large bowl. Blend together honey, mayonnaise, salt, and cream and fold in nuts. Pour honey mixture over apple mixture, gently toss, and refrigerate before serving.

Company Coleslaw

6 cups (420 g) cabbage, shredded

2 cups (240 g) carrots, shredded

1 green pepper, chopped

½ cup (170 g) honey

½ cup (120 g) mayonnaise

½ cup (120 ml) white vinegar

½ tsp. salt

¼ tsp. pepper

2 tbsp. (28 ml) fresh lemon juice

1 tsp. celery seed

Mix together cabbage and carrots. Mix in green pepper. Blend the rest of the ingredients together in a separate bowl and pour over cabbage and carrots. Gently toss, cover, and refrigerate overnight. Serves six to eight.

Carrot Raisin Salad

3 cups (360 g) raw carrots, shredded

1 cup (165 g) seedless raisins

⅓ cup (75 g) mayonnaise or salad dressing

¼ cup (60 ml) milk

1 tbsp. (20 g) honey

1 tbsp. (15 ml) lemon juice

¼ tsp. salt

Toss carrots with raisins. Blend together remaining ingredients to make dressing and mix into carrot-raisin mixture. Refrigerate for about 30 minutes. Makes eight servings.

desserts

Scott Scones

1 cup (125 g) flour
dash salt
2 tsp. baking powder
3 tbsp. (42 g) butter
1 egg
3 tbsp. (60 g) honey
¼ cup (42 g) raisins
milk

Mix together flour, salt, and baking powder. Add butter and mix with a fork until crumbly. Beat the egg, and to it add the honey. Add this to the flour mixture. Then add the raisins. Form the dough into a ball. Flour a cutting board. Roll out the dough with a floured rolling pin to about ½" (1.3 cm) thick. Cut into circles with cutter or top of a water glass. Brush the tops of each scone with milk. Grease cookie sheet, and place the scones on it 1" (2.5 cm) apart. Bake 10 minutes in 425° F (218° C) oven. Do not overbake.

Crispy Honey Cookies

1 6-ounce (170-g) package chocolate bits
dash salt
¼ cup (85 g) honey, light
1 tbsp. (15 ml) water
1 tsp. (5 ml) vanilla
3 cups (75 g) crisped rice cereal

Mix together all ingredients except cereal in a saucepan. Over very low heat, blend and stir until smooth and creamy. Pour this syrup over the cereal, stir and toss lightly to coat. Spread and pat firmly into place in a buttered 8" × 8" × 2" (20 × 20 × 5 cm) glass baking dish. Cool before cutting. Makes two dozen squares.

Honeyscotch Sundae

6 tbsp. (84 g) butter
2 tsp. cornstarch
1⅓ cups (454 g) honey
½ cup (65 g) chopped pecans
vanilla ice cream

Melt butter over low heat; stir in cornstarch. Add honey and cook, stirring constantly until mixture boils. Add pecans. Serve warm over ice cream. Makes about 1½ cups (355 ml).

Perfect Pound Cake

1 cup (225 g) butter
1 cup (340 g) honey
4 eggs, beaten
1 tsp. (5 ml) vanilla
3 cups (375 g) cake flour
3 tsp. baking powder
¼ tsp. salt
1 tsp. (5 ml) almond flavoring

Cream the honey and butter. Add the beaten eggs and vanilla. Sift together the dry ingredients, then stir into honey-egg mixture. Beat well. Add almond flavoring, pour into a greased loaf pan, and bake in 300° F (149° C) oven for 2 hours. You can vary this recipe by adding a cup (165 g) of raisins or currants or ½ cup (65 g) chopped pecans.

Honey Zucchini Bread

1 egg

¾ cup (255 g) honey

3 tbsp. (45 ml) vegetable oil

1 tsp. (5 ml) vanilla

2¼ cups (313 g) all-purpose flour

1½ tsp. baking powder

1 tsp. orange peel, grated

½ tsp. baking soda

½ tsp. ground ginger

¼ tsp. salt

1¼ cups (180 g) zucchini, grated

Beat egg slightly in a large bowl, then add honey, oil, and vanilla and mix well. Combine flour, baking powder, orange peel, baking soda, ginger, and salt in medium bowl. Add flour mixture and zucchini to the honey mixture, and mix until blended. Spoon batter into well-greased 9" × 5" × 3" (22.5 × 12.5 × 7.5-cm) loaf pan. Bake at 325° F (163° C) about 1 hour or until wooden toothpick comes out clean. Cool 20 minutes in pan, then remove from pan.

Zucchini Chocolate Cake

⅔ cup (150 g) butter

¾ cup (255 g) honey, light

¾ cup (170 g) packed brown sugar

3 eggs

1 tsp. (5 ml) vanilla

2¼ cups (313 g) flour

1 tsp. baking powder

1 tsp. baking soda

¼ tsp. salt

½ cup (50 g) unsweetened cocoa

1 cup (235 ml) milk

1 cup (120 g) zucchini, grated

Cream butter, honey, and sugar. Beat in eggs and vanilla. Sift together the dry ingredients and gradually stir into honey mixture, alternating with milk. Finally, fold in the grated zucchini and pour into round layer cake pans. Bake at 350° F (177° C) for 30–35 minutes.

Chocolate Fudge

⅔ cup (227 g) honey
2 cups (400 g) sugar
¾ cup (175 ml) heavy cream
1 square unsweetened chocolate, grated
1 tsp. (5 ml) vanilla
½ cup (63 g) chopped walnuts
¼ tsp. salt

In a medium saucepan, mix honey and sugar together. To that, add cream and grated chocolate. Mix well. Place this mixture over low heat and cook slowly, stirring until no grains of sugar can be felt. Then, increase the heat and cook until a few drops form a medium-firm ball when dropped into cold water. Remove from the heat, add salt and vanilla, and beat until very thick and the mixture loses its shiny gloss. This step takes about 20 minutes, so plan accordingly. Add the nuts just before pouring the mixture into a buttered pan. When cool and solid, cut into squares.

JUST FOR FUN

Bar-Le-Duc Preserves

Read this, and try it at least once.

These preserves are believed to be the finest of their kind, and have hitherto been imported at extravagant prices. Other fruits besides currants may be treated in this way, because honey is a preservative. These preserves do not need to be kept absolutely airtight.

Take selected large red or white currants. One by one, carefully make an incision in the skin ¼" (.6 cm) deep with tiny embroidery scissors. Through this slit remove the seeds with the aid of a sharp needle, preserving the shape of the fruit. Heat honey (use the same amount as the weight of the currants). When warm, add the currants. Let it simmer 1–2 minutes, and then seal in jelly jars. The currants retain their shape, are a beautiful color, and melt in the mouth. Care should be exercised not to scorch the honey.

Herbal Honey

You can add a hint of fresh herb flavor to your light and medium honeys. First, gather the herbs as early in the morning as possible while they're still fresh and turgid (not wilted) and full of flavor. Bring into the kitchen and wash thoroughly under cold water to remove soil, insects, spent flowers, or old leaves. Almost any herb can be used. Consider delicate-flavored herbs such as rose petals, chamomile, or lavender, or stronger herbs, such as rosemary, anise, or any of the many mints.

Use a light, mildly flavored spring honey. Once you've filtered your crop, fill pint jars about ⅘ full so you can add and remove the herbs without overflowing the honey. At a local craft, hobby, or organic food store, purchase some oversize reusable tea bags or make your own using fine mesh nylon from a fabric store. The bags should be 2"–4" square (5–10 cm square) with either a drawstring or other clamp on the opening.

Chop delicate-flavored herbs into moderate to fine pieces and add 3–5 tbsp. to your mesh bag and seal. For stronger herbs, use 2–4 tbsp.

Place a bag in each pint jar and set in a sunny windowsill for 1–2 weeks. After the first week or so, taste the honey, and if strong enough, remove the bag. If more flavor is desired, either add more herbs or allow the jar to sit in the sun for another week. After the flavor has reached its peak, remove the honey from the windowsill, carefully remove the bag of herbs, and discard it. You can then rebottle the honey in fancy gift jars for friends and family, or to sell.

If the honey becomes too strongly flavored, simply dilute with unflavored honey and mix well.

If, no matter what you do, you can't detect the herbal flavor, try the following: Obtain more herbs, chop them in smaller pieces, and increase the amount you add to the bag. Then, loosen but do not remove the cover and place herbs and honey in your oven at about 120° F (49° C) for 4–6 hours. If after this treatment your honey remains unflavored, choose another herb.

Use your herb-flavored honey in any of the recipes here where the herb would add to the distinction and flavor of the final product. You can also use this simply as table honey. If you want, combine several herbs in the same jar for a unique or exotic blend.

Glossary

A.I. Root—founder of the first and once the largest beekeeping equipment manufacturing company in the U.S., located in Medina, Ohio.

Abdomen—the third region of a body of a bee enclosing the honey stomach, intestine, reproductive and other organs, wax and Nasonov glands, and the sting.

Abscond—the action of all bees leaving the hive due to extreme stress, disease, pests, or danger, such as a fire.

African honey bees—a subrace of honey bees, originally from Africa, brought to Brazil, that has migrated north to the U. S. They are extremely defensive and nearly impossible to work.

Alarm pheromone—pheromone released by worker bees during an emergency.

American foulbrood (AFB)—a brood disease of honey bees caused by the spore-forming bacterium *Paenibacillus* (formerly *Bacillus*) *larvae*.

Anther—the part of a flower that produces pollen; the male reproductive cells.

Apiary—where honey bee colonies are located; often called beeyard.

Apiculture—the science and art of keeping honey bees.

Apis mellifera—the genus and species of the honey bee found in the United States.

Bait hive—a box, often an old brood box, composed of a comb or two, a top and bottom, and a small entrance hole, used to attract swarms. It is often placed in an apiary.

Balling the Queen—The action of worker bees attacking a new queen, or a queen cage, intent on killing her because she is foreign. Often occurs during queen introduction.

Bee bread—a mixture of pollen and honey used as food by the bees.

Bee escape—a device used to remove bees from honey supers during harvest by permitting bees to pass one way but preventing their return.

Bee space—¼" – ⅜" space that bees live in.

Beeswax—a complex mixture or organic compounds secreted by eight glands on the ventral side of the worker bee's abdomen; used for molding six-sided cells into comb. Its melting point is from 144° F (62° C) to 147° F (64° C).

Bee veil—a cloth or wire netting for protecting a beekeeper's face, head, and neck from stings.

Bee venom—the poison secreted by glands attached to a bee's stinger.

Bottom board—the screened floor of a beehive.

Brace/burr comb—comb built between parallel combs, adjacent wood, or two wooden parts such as top bars.

Brood—the term used for all immature stages of bees: eggs, larvae, and pupae.

Brood chamber—the part of the hive in which the brood is reared.

Capped brood—pupae whose cells have been sealed as a cover during their nonfeeding pupal period.

Cappings—the thin, pure wax covering of cells filled with honey; the coverings after they are sliced from the surface of a honey-filled comb when extracting the best beeswax.

Carniolan—dark-colored race of bees from Eastern Europe, which are very gentle.

Caucasian—grayish-colored race of bees from Europe, use excessive propolis.

Cell—a single hexagonal (six-sided) compartment of a honey comb.

Chalkbrood—a fungal disease of honey bee larvae.

Chilled brood—developing bee brood that have died from exposure to cold.

Cleansing flight—a quick, short flight bees take after confinement to void feces.

Cluster—a group of bees hanging together for warmth.

Colony—adult bees and developing brood living together including the hive they are living in.

Comb—a sheet of six-sided cells made of beeswax by honey bees in which brood is reared and honey and pollen are stored.

Comb foundation—a commercially made sheet of plastic or beeswax with the cell bases of worker or drone cells embossed on both sides.

Comb honey—honey produced and sold in the comb, made in plastic frames and sold in round, plastic packages.

Compound eyes—a bee's sight organs, which are composed of many smaller units called ommatidia.

Cremed (Crystallized) honey—honey that has been allowed to crystallize under controlled conditions.

Dancing—a series of repeated movements of bees on comb used to communicate the location of food sources and potential home sites.

Dearth—a time when nectar or pollen or both are not available.

Dividing—partitioning a colony to form two or more units, often called divides or splits.

Drawn comb—comb with cells built out by bees from a foundation.

Drifting—bees going to a colony that is not their own.

Drone—the male honey bee.

Drone comb—comb measuring about four cells per inch in which the queen lays unfertilized eggs that become drones.

Drone layer—a queen able to produce only unfertilized eggs, thus drones.

Egg—the first stage of a honey bee's metamorphosis.

Entrance reducer—a wooden or metal device used to reduce the large entrance of a hive to keep robbing bees out and to make the entrance easier to defend, and to reduce exposure to wind and the elements outside.

European foulbrood (EFB)—an infectious brood disease of honey bees caused by the bacterium *Melissococcus* (formally *Streptococcus*) *pluton*.

Extracted honey—liquid honey removed from the comb with an extractor.

Fanning or scenting—worker bees producing Nasanov pheromone and sending it out to bees away from the colony as a homing beacon.

Feeder—any one of a number of devices used to feed honey bees sugar syrup including pail feeders, inhive frame feeders, hive-top feeders, and entrance feeders.

Fertile queen—a queen that can lay fertilized eggs.

Forager—worker bees that work (forage) outside the hive, collecting nectar, pollen, water, and propolis.

Frame—four pieces of wood/plastic (top bar, a bottom bar, and two end bars) designed to hold foundation/drawn comb.

Fumidil-B—one trade name for fumagillin; a chemotherapy used in the prevention and suppression of Nosema disease.

Fume board—a rectangular frame, the dimensions of a super, covered with an absorbent material such as cloth or cardboard, on which a chemical repellent (Bee Go or Bee-Quick) is placed to drive the bees out of supers for honey removal.

Granulation—the formation of sugar (glucose) crystals in honey.

Grease patty—a mixture of vegetable shortening and granulated sugar placed near the brood area for tracheal mite control.

Hive—a man-made home for bees.

Hive tool—a metal tool used to open hives, pry frames apart, and remove wax and propolis.

Honey—a sweet material produced by bees from the nectar of flowers, composed of glucose and fructose sugars dissolved in about 18 percent water; contains small amounts of sucrose, mineral matter, vitamins, proteins, and enzymes.

Honey flow—a time when nectar is available and bees make and store honey.

Honey stomach—a portion of the digestive system in the abdomen of the adult honey bee used for carrying nectar, honey, or water.

Hymenoptera—the order of insects that all bees belong to, as do ants, wasps, and sawflies.

Inner cover—a lightweight cover used under a standard telescoping cover on a beehive.

Italian bees—most widely used race of honey bees in the United States; originally from Italy.

Langstroth hive—our modern-day, man-made, moveable frame hive named for the original designer.

Larva (plural, larvae)—the second (feeding) stage of bee metamorphosis; a white, legless, grublike insect.

Laying worker—a worker that lays drone eggs, usually in colonies that are hopelessly queenless.

Marked queen—Some queen producers sell queens that they mark with a spot of paint on the top surface of the thorax (the middle of three chief divisions of an insect's body). This makes the queen much easier to find, and indicates whether the queen you have found is the one you introduced or a new queen. Always use marked queens.

Mating flight—the flight made by a virgin queen when she mates in the air with several drones.

Metamorphosis—the four stages (egg, larva, pupa, adult) through which a bee passes during its life.

Nectar—a sweet liquid secreted by the nectaries of plants to attract insects.

Nosema—A digestive disease of honey bees, treated with Fumigillin.

Nuc or Nucleus (plural, nuclei)—a small, two- to five-frame hive used primarily for starting new colonies.

Nurse bees—young bees, three to ten days old, that feed and take care of developing brood.

Ocellus—simple eyes (3) on top of a honey bee's head. Used primarily as light sensors.

Package bees—screened shipping cage containing three pounds of bees, usually a queen, and food.

PDB (paradichlorobenzene)—crystals used as a last resort as a fumigant to protect stored drawn combs against wax moth.

Pheromone—a chemical secreted by one bee that stimulates behavior in another bee. The best known bee pheromone is queen substance secreted by queens that regulate many behaviors in the hive.

Pollen—the male reproductive cells produced by flowers, used by honey bees as their source of protein.

Pollen basket—a flattened area on the outer surface of a worker bee's hind legs with curved spines used for carrying pollen or propolis to the hive.

Pollen trap—a mechanical device used to remove pollen loads from the pollen baskets of returning bees.

Pollination—the transfer of pollen from the anthers to the stigma of flowers.

Proboscis—the mouth parts of the bee that form the sucking tube and tongue.

Propolis—sap or resinous materials collected from the buds and wounds of plants by bees, then mixed with enzymes and used to strengthen wax comb, seal cracks and reduce entrances, and smooth rough spots in the hive.

Pupa—the third stage in the metamorphosis of the honey bee, during which the larva goes from grub to adult.

Queen—a fully developed female bee capable of reproduction and pheromone production. Larger than worker bees.

Queen cage—a small cage used for shipping and/or introduction of a queen into a colony.

Queen cell—a special elongated cell, in which the queen is reared. Usually an inch or more long, has an inside diameter of about ⅓ inch, and hangs down from the comb in a vertical position, either between frames or from the bottom of a frame.

Queen cell cup—A round, cup-shaped structure that workers build on the bottoms of frames to accommodate a future queen cell. The current queen must place an egg in the cup before the workers begin building the rest of the queen cell. Queen cell cups are built most often just before swarm behaviors begin.

Queen excluder—metal or plastic grid that permits the passage of workers but restricts the movement of drones and queens to a specific part of the hive.

Queenright—a colony with healthy queen.

Rabbet—a narrow ledge on the inside upper end of a hive body or super from which the frames are suspended.

Requeen—to replace existing queen with new queen.

Robbing—bees stealing honey, especially during a dearth, and generally from weaker colonies.

Royal jelly—a highly nutritious glandular secretion of young bees, used to feed the queen and young brood.

Scout bees—foraging bees, primarily searching for pollen, nectar, propolis, water, or a new home.

Small hive beetle—a destructive beetle that is a beehive/honey house pest living generally in the warmer areas of the U.S. Originally from South Africa.

Smoker—a device used to produce smoke, used when working a colony.

Solar wax melter—a glass-covered insulated box used to melt wax from combs and cappings.

Spermatheca—an internal organ of the queen that stores the sperm of the drone.

Sting—the modified ovipositor of a honey bee used by workers in defense of the hive and by the queen to kill rival queens.

Sucrose—principal sugar found in nectar.

Super—a hive body used for storing surplus honey placed above the brood chamber.

Supersedure—a natural or emergency replacement of an established queen by a daughter in the same hive.

Surplus honey—honey stored by bees in the hive that can be used by the beekeeper and is not needed by the bees.

Swarm—about half the workers, a few drones, and usually the queen that leave the parent colony to establish a new colony.

Swarm cell—developing queen cell usually found on the bottom of the frames reared by bees before swarming.

Terramycin—an antibiotic used to treat European foulbrood. Also used for American foulbrood prevention, but it is not effective in killing the spore stage of this disease.

Thorax—the middle section of a honey bee, that has the wings and legs and most of the muscles.

Tracheal mite—*Acarapis woodi*, the tracheal-infesting honey bee parasite.

Tylosin—one of several antibiotics used to treat, but not cure, American foulbrood.

Uncapping knife—a specially shaped knife used to remove the cappings from sealed honey.

Uniting—combining two or more colonies to form a larger colony.

Varroa mite—*Varroa destructor*, a parasitic mite of adult and pupal stages of honey bees.

Venom—the chemical injected into the skin when a honey bee stings. It's what makes being stung painful.

Virgin queen—an unmated queen.

Wax moth—larvae of the moth *Galleria mellonella*, which damages brood combs.

Worker bee—a female bee whose reproductive organs are undeveloped. Worker bees do all the work in the colony except for laying fertile eggs.

Worker comb—comb measuring about five cells to the inch in which workers are reared.

✦Resources✦

Books

The ABC & XYZ of Beekeeping, 41st Edition
Ed. by H. Shimanuki, Ann Harman
A.I. Root Co.

Africanized Honey Bees in the Americas
Dewey Caron
A.I. Root Co.

Aromatherapy Creams and Lotions
Donna Maria
Storey Books

The Bee Book
Beekeeping in the Warmer Areas of Australia
Warhust & Goebel
DPI, Queensland

The Beekeeper's Garden
Hooper & Taylor
A&C Black Plc, London

The Beekeeper's Handbook
Sammataro & Avitabile
Cornell University Press

Beekeeping for Dummies
Howland Blackiston
Hungry Minds Press

Beekeeping Principles
James E. Tew
W. T. Kelley

Beeswax Crafting
Robert Berthold Jr.
Wicwas Press

The Biology of the Honey Bee
Mark Winston
Harvard University Press

A Book of Bees, and How to Keep Them
Sue Hubbell
Random House

Candle Makers' Companion
Betty Oppenheimer
Storey Books

Control of Varroa for New Zealand Beekeepers
Goodwin and van Eaton
New Zealand Ministry of Agriculture

Covered in Honey, Cookbook
Mani Neall
Rodale Press

Dance Language of the Bees
Karl von Frisch
Harvard University Press

An Eyewitness Account of Early American Beekeeping
A.I. Root
A.I. Root Co.

Following the Bloom
Douglas Whynott
G.P. Putnam

The Forgotten Pollinators
Buchman and Nabor
Island Press

Form and Function in the Honey Bee
Lesley Goodman
IBRA

The Hive and the Honey Bee
Ed. By Joe Graham
Dadant & Sons

Honey Bee Pests, Predators, and Diseases
Ed. By Morse and Flottum
A.I. Root Co.

A Honey Cookbook
Recipes from the Home of the Honey Bee
A.I. Root Co.

Honey Plants of North America (reprint of 1926 Edition)
Harvey Lovell
A.I. Root Co.

Honey the Gourmet Medicine
Joe Traynor
Kovak Books

The How-to-Do-It Book of Beekeeping
Richard Taylor
Ross Rounds, Inc.

The Joys of Beekeeping
Richard Taylor
Linden Press

Langstroth's Hive and the Honey Bee (reprint. 4th Edition)
L.L. Langstroth
Dover Books

Making Wild Wines and Meads
Vargas and Guilling
Storey Books

Queen Rearing and Bee Breeding
Laidlaw and Page
Wicwas Press

Observation Hives
Webster and Caron
A.I. Root Co.

Practical Beekeeping in New Zealand
Andrew Matheson
GP Publications, Wellington

The Sacred Bee
Hilda Ransom
Dover Books

Soap Makers' Companion
Susan Cavitch
Storey Books

Weeds of the Northeast
Uva, Neal and Thomas
Cornell University Press

Weeds of the West
Ed. by University of Wyoming
Western Society of Weed Science

What Do You Know?
Clarence Collison
A.I. Root Co.

The Wisdom of the Hive
Thomas D. Seeley
Harvard University Press

The World History of Beekeeping and Honey Hunting
Eva Crane
IBRA

Magazines

North America

American Bee Journal
Dadant & Sons
51 S. 2nd Street
Hamilton, IL 62341
Phone: 1.217.847.3324
www.dadant.com

Bee Culture, The Magazine of American Beekeeping
A.I. Root Co.
623 W. Liberty Street
Medina, OH 44256
Phone: 800.289.7668
www.BeeCulture.com
This Web page is the source of information for beekeeping associations and state inspection agencies.

Hive Lights
Canadian Honey Council
234-5149 Country Hills Blvd. NW, Ste. 236
Calgary, AB T3A 5K8
Canada
www.honeycouncil.ca

Australia and New Zealand

The Australasian Beekeeper
34 Racecourse Road
Rutherford, NSW 2320
Australia
penders@nobbys.net.au

The New Zealand Beekeeper
48 Stafford Street
P.O. Box 5002
Dunedin
New Zealand
www.nba.org.nz

Europe

An Beachaire
The Irish Beekeeper
Weston, 38 Elton Park
Sandycove, Co. Dublin
Ireland
www.irishbeekeeping.ie

Bee Craft
107 Church St., Werrington,
Peterbourough PE4 6QF
United Kingdom
www.bee-craft.com

The Beekeepers' Quarterly
Northern Bee Books
Scout Bottom Farm
Mytholmroyd, Hebden Bridge
West Yorkshire HX7 5JS
United Kingdon
Jeremy@recordermail.demon.co.uk

International Bee Research Association
Bee World, and *Journal of Apicultural Research*
18 North Rd.
Cardiff, CF10 3DT
United Kingdom
Phone: 44.0.29.2037.2409
www.IBRA.org.uk

Norges Birokteren
Bergerveien 15
N-1396 Billingstad
Norway
www.norges-birokterlag.no

Redaktion Deutsches Bienen-Journal
Postfach 31 04 48
10634
Berlin
Germany
www.bienenjournal.de/bienen/

La Sante De L'Abeille
Federation Nationale
Des Organisations Sanitaires Apicoles Departementales
41, rue Pernety
75014 Paris
France
Phone: 33.0.4.92.77.75.72
www.sante-de-labeille.com

Vida Apicola
Ausias March,
25, 1° 08010
Barcelona
Spain
www.vidaapicola.com

Suppliers

B&B Honey Farm (general supplies)
5917 Hop Hollow Rd.
Houston, MN 55943
Phone: 507.896.3955
www.bbhoneyfarm.com

Mann Lake, Ltd. (general supplies)
501 S. First Street
Hackensack, MN 56452-2001
Phone: 800.880.7694
www.mannlakeltd.com

Ross Rounds
P.O. Box 11583
Albany, NY 12211
Phone: 518.370.4989
www.rossrounds.com

Dadant & Sons (general supplies)
51 S. 2nd St.
Hamilton, IL 62341
Phone: 217.847.3324
www.dadant.com

Betterbee, Inc. (general supplies)
8 Meader Road
Greenwich, NY 12834
Phone: 800.632.3379
www.betterbee.com

Brushy Mountain Bee Farm (general supplies)
610 Bethany Church Road
Moravian Falls, NC 28654
Phone: 800.233.7929
www.beeequipment.com

Kelley's Bee Supply (general supplies)
807 W Main Street
Clarkson, KY 42726
Phone: 270.242.2012
www.kelleybees.com

E.H. Thorne, Ltd. (general supplies)
Beehive Works, Wragby
Market Rasen LN8 DLA
United Kingdom
www.thorne.co.uk

Pender's Beekeeping Supplies (general supplies)
28 Munibung Road
Cardiff NSW 2285
Australia

Rossman Apiaries
P.O. Box 909
Moultrie, GA 31776
Phone: 800.333.7677
www.GaBees.com

⟢Index⟣

⇥Illustration Credits⇤

All photographs by Kim Flottum with the exception of the following pages:

Courtesy of A. I. Root, 10; 34; 37; 38; 98 (top & bottom, left); 99 (top)
Robin Bath, 113; 117; 120; 125; 126; 127; 128; 129; 130; 132; 136; 145; 149; 153
R. Chamberlin, 33; 59 (top)
E. R. Jaycox, 8, 45 (bottom); 53; 78 (bottom); 83 (bottom right)
Serge Labesque, 91
Ohio Department of Agriculture, 92 (bottom)
Allan Penn, 116; 124; 141; 143
Dr. James E Tew, 34
Courtesy of United States Department of Agriculture, 55 (right); 85 (top)
Jennifer Wills, 138; 148
Michael G. Yatcko, illustrations, 7 (top, bottom right); 15; 18; 32; 41; 49; 54; 62; 68 (top)

About the Author

After receiving a degree in production horticulture from the University of Wisconsin, Kim Flottum began a career in honey bee pollination research with the United States Department of Agriculture and a lifelong interest in the multifaceted hobby and business of beekeeping. He next used his acquired skills to raise apples and vegetables in Connecticut, before moving to Medina, Ohio, in 1986 to become editor of the 132-year-old magazine, *Bee Culture*, where he has remained for over 18 years.

A full schedule of lectures, magazine deadlines, and volunteer work in beekeeping organizations including the Eastern Apiculture Society, which he chairs, and the Medina County Beekeeper's Association, of which he is president, infringe on weekend leisure time, but Kathy Summers, his partner in beekeeping, helps with their two backyard honey bee colonies and a small garden that offers occasional salads and a summer-long supply of flowers (and weeds).

Kim Flottum brings a wealth of experience in gardening, beekeeping, and communications together in *Backyard Beekeeping* and lives by the Beekeeper's credo: *Next year will be better!*